The Complete Guide to Building Your Own 8-inch Telescope

By Kevin J Manning

I've written a book on how to build a telescope just like the one I did so anyone can do it. Every step and every piece is covered in great detail so you can actually build a powerful telescope much more powerful than the one's typically found in department stores. The optics are larger with greater light-gathering power and resolution, enabling you to observe fainter stars, nebulae, and galaxies, and to see finer detail on planets. The high magnifications advertised with department store telescopes are actually useable with this telescope. The eyepieces are larger with a bigger window for the eye to look into, not something you must squint and strain to see anything. The mounting is very rigid and stable, not like the flimsy tripods that move and vibrate to the point you have to move your head around trying to keep up with the bouncing image. The simple design makes the instrument able to be set up in seconds anywhere you bring it. If you follow the instructions in the order given it's almost impossible to fail. The large optics gather 500 times more light than the unaided eye, permitting stunning views of the universe. Amaze your family and friends with the achievement of making a powerful telescope that will yield a lifetime of beautiful views of the heavens. This would make a great parent-child project that would prove very rewarding. Be prepared for sighting celestial objects like you've never seen them before with this truly powerful 8-inch reflecting telescope.

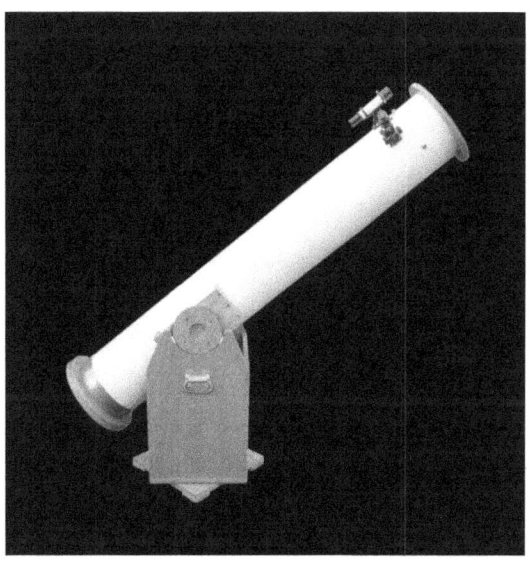

Table of Contents

This book is dedicated to God, who created the heavens and the earth, and the memory of my parents, Walter and Margaret Manning, who not only instilled in me an idealistic world view full of promise at an early age, but also gifted me with my first two telescopes to begin my journey of exploring the heavens.

Preface

Psalm 19:1 records that "The heavens declare the glory of God; and the firmament sheweth his handywork." The beauty and majesty of a starry sky is indeed awe-inspiring to the beholder, and one must look up to see and wonder at its splendor. The universe has always been my laboratory containing untold mysteries in the depths of space. The jewels of the night sky march across this panorama with smooth and gracious transitions as the earth spins without compromise. Day after day we are reminded that time is artificially induced and the patience of the cosmos commands our attention to what is really important.

Familiar figures in the sky take shape connecting stars of a constellation. Our unaided eyes alone reveal the dim glow of the Milky Way stretching across our view, while streaks of light from a meteor shower require nothing more than our eyes and attention. As they become more adapted, glowing clouds of gas and dust along with glittering clusters of stars like diamonds set in black velvet begin to emerge out of the darkness. The serene and calm appearance of the night sky can be misleading as we realize the dangerous and abundant violence throughout the universe in the form of exploding stars, intense radiation, and extreme cold. They beckon us to inspect them more closely.

Even a small pair of binoculars will reveal hundreds of stars in the seeming void of dark space. Striking colors inhabiting the rainbow are seen in these tiny points of light. In the deep regions of the cosmic ocean there are things that defy imagination just waiting for their story to be told in the light path of a telescope. The ice crystal rings of Saturn look unreal and the dance of Jupiter's four brightest moons remind us of Galileo's first view of them four hundred years ago. The craters upon the lunar landscape are truly breathtaking, while faint fuzzy patches of light inspire curiosity.

Larger telescopes gather more light, allowing us to see even deeper into space. Remote star clusters are resolved through their core, each with distinction, and distant suns show evidence of both their birthplace and deathbed in ghostly swaths of gas and dust, the colors of these telling of their chemical composition. Very far-away "island universes" called galaxies come in a variety of shapes and sizes within gigantic clusters themselves, each containing hundreds of billions of stars, gas and dust.

Astronomy is truly an incredible experience for young and old alike. You are put into the role of a discoverer, seeking out new places never seen before, and are able to venture into realms you could never go to from your own backyard. To build such an instrument that would take us there without even leaving the ground is a joy unspeakable and never to be forgotten.

Introduction

Imagine you're a child again. Your parents drove you with them on a vacation far away in a quiet secluded spot away from civilization, as you knew it back home. As you exit off the small country road onto the grassy path leading up to an old two-story log cabin with a big front porch, your stomach begins to growl as you anticipate the delicious foods brought along for the cookout. After emptying the car of the gear and eating an early dinner, you relax on the front porch swing. The only sounds you hear are the occasional flocks of birds flying overhead and the gentle breeze. You watch as a beautiful sunset appears in the western sky and enjoy the peaceful solitude and surroundings. Just before darkness sets in, you take a short walk down by the lake and lay down in the grass. As nightfall grows darker you begin to notice that more and more stars appear in the sky. Like glistening diamonds set in black velvet, so many more stars blanket the sky than you were used to seeing back home that you begin feeling a sense of awe and wonder.

A little while later you hear the voices of your dad and sister approaching. They have red filtered flashlights so it doesn't disturb your night vision. As they draw near, you notice that they are carrying the telescope all of you recently completed building back home, but never looked through except to align the optical system. You jump up full of excitement and ask if you can point the telescope toward this really bright star that doesn't appear to twinkle like the others. You begin to gasp saying, "Wow" as you focus on this bright yellow ball surrounded by equally bright rings tilted in the field of view. As you begin to gain some composure you announce, "It looks like the planet Saturn I saw in a book once," your sister and dad begin begging for a look. Your dad recalls reading that Saturn is nearly one billion miles away, is the second largest planet far exceeding the size of the earth, and has a density low enough to float on water. Your previous sense of awe and wonder is rekindled once again.

As dad grows tired he returns to the cabin leaving you and your sister with the telescope. As the two of you scan across the crystal clear sky, every so often you "bump" into clusters of stars appearing as jewels, some having different colors, and large patches of light with different shapes (not knowing they are gigantic clouds of glowing gas and dust called nebulae and remote distant galaxies containing hundreds of billions of stars in their own separate "island universe"). Once again, you experience a sense of awe and wonder.

After a while your sister becomes tired and leaves you alone with the telescope. You gaze over toward the east and notice the Moon rising in the sky, and quickly turn the telescope to point in its direction. You find the Moon's image easily because of its large size and brightness. As you finely tune the focus you can barely keep yourself from screaming with excitement as you view numerous craters amidst a wealth of detail almost beyond description. Now your sense of awe and wonder has made an indelible impression, and even after you pack it in returning to the cabin for sleep, you begin to dream of the great celestial timepiece turning overhead filled with uncountable new discoveries.

Why should anyone spend the time and effort to build their own telescope? Here are a few good reasons:

- For many years, department stores have been selling small refractor telescopes using false advertising of their product's performance capability. Though it is true that an extended range of magnifications can be achieved with any telescope using various eyepieces and Barlow lenses, which does not at all mean the telescope can handle them. When these small department store telescopes are advertised as 540X or 600X, the consumer is misled in two ways. First, a telescope's "power" is not determined by the amount of magnification (the number of times the diameter of an object appears to be larger) used on an object's image. When it comes to telescopes, the old cliché "bigger is better" is absolutely true. Of the five main functions of a well-performing telescope: collect sufficient light, adequate resolution, good definition, magnification, and field of view, the primary function is that of light-gathering power. Larger diameter lenses or mirrors (commonly referred as aperture), yielding more surface area for light collection and known as the primary objective (the heart of the telescope), is what produces a brightly illuminated image for viewing. Another function associated directly with objective size is resolving power, which is the ability to separate objects that appear close together and distinguish fine details on extended objects, such as the moon and planets. The larger the objective, the more "powerful" the telescope is to detect faint, distant objects deeper into space, and to display details in those objects. Second, there are limits to useful magnification. Atmospheric turbulence is one factor, and size of the objective is another. Under ideal conditions, an upper limit of 60X per inch of aperture is a good rule of thumb. The most common telescope sold at department stores typically has an objective lens diameter of 60 millimeters, or 2.4 inches. Using our formula, 2.4 x 60 = 144X. In other words, this is the highest useful magnification possible under ideal atmospheric conditions. Magnifications beyond this are not useful because the image deteriorates and eventually becomes a blur. This situation is commonly known as "empty magnification." A telescope able to handle the advertised magnifications would need an objective diameter of 10 inches or larger.

- Most of these department store telescopes utilize small Japanese eyepieces with barrel diameters of 0.965 inches, many which have extremely small lenses that the human eye is barely able to look through. This usually creates a painful and dissatisfying viewing experience for the user. U.S. standard size eyepieces use 1.25 inch diameter barrels, and lenses typically wide enough to enjoy a more comfortable view. There are even eyepieces with 2-inch diameter barrels readily available. These often employ lenses for the eye big enough for views like looking through a window to the universe. When you build your own telescope, you get to choose the size of the eyepieces, and select those giving a range of magnifications that are useful and practical.

- The mounting that supports the telescope's optical tube assembly that you will be building is a basic Dobsonian design that is far more stable than the flimsy mountings and tripods often used in department store telescopes. This type of mounting is basically indestructible and very simple to use for everyone. It is also easy to transport and set up. You simply place the mounting anywhere on the ground you find a relatively flat and level surface and then place the optical tube assembly on top. It literally only takes a few seconds to do.

- The inherent joy and satisfaction one feels when constructing their own optical instrument is irreplaceable. Especially something built to last a lifetime and handcrafted to function impeccably well. Once completed and ready for use, the exhilaration of the first time you set up the telescope and experience "first light" is something never to be forgotten.

- A project such as building a telescope is a lot of fun that can be shared between a father and son or even the entire family. The special occasion of working together toward the same goal brings much joy and fulfillment. This unforgettable experience is only enhanced by the years of enjoyment shared while using the telescope to explore the heavens each available clear night.

In essence, the journey of building your own telescope can only be outweighed by the lifetime of using such a time machine to explore the hidden mysteries of the universe.

Telescopes 101

The clear night sky has always presented spectacular views to the unaided eye. However, even the slightest telescopic aid will reveal much more. Not only are we able to detect what was beforehand "invisible" to the unaided eye, we are also looking further back in time as we probe deeper into the universe. This is because the distance of many objects we can see in space are so vast that it takes time, often many hundreds, thousands, millions, even billions of years for its light to reach us on earth. The question is in what way is this great feat accomplished and to what degree?

The word telescope literally is defined as an instrument used "to see distant objects." It is the primary tool of the astronomer. In its most basic form, there are two types of telescopes – refractor and reflector. As mentioned in the introduction, the primary objective is the heart of the telescope that must gather sufficient light from an object to produce a brightly illuminated image and have sufficient resolving power to separate close-together objects that appear as one to the unaided eye. The refracting telescope uses an achromatic objective lens as its primary light collector. It is called a refractor because the curved lens refracts, or bends light to bring it to a focus. Galileo Galilei was the first to use a refractor telescope for astronomical purposes. You may recall putting coins in a refractor telescope mounted by the edge of a tall building or a large body of water to help see objects that are far away. Binoculars are essentially two refractor telescopes mounted side by side enabling us to use both eyes simultaneously.

7x35 binoculars means that each of the primary objective lenses in front are 35 millimeters (about 1.4 inches) in diameter, and that they magnify images to appear 7 times larger in diameter compared with the view using unaided eyes alone. 10x50 binoculars magnify 10 times and have lenses nearly 2 inches across. At night, it's not so much the difference in magnification that makes the 10x50's more powerful, but the larger area of the 50 mm primary lenses collect more light than the 35 mm lenses do. Have you ever noticed the pupils of your eyes getting smaller when walking into a dark room and switching on the light? Without even thinking about it, your eyes automatically "adjust" to the lighting conditions present. For the same reason, you are able to see more and more in time after leaving a brightly lit house to go outdoors at night. After about 20 minutes, our pupils enlarge to their widest opening, around 7 mm (about ¼ inch) for a young person. The rods in our eyes are sensitive to faint light, and chemical reactions assist further in our eyes becoming dark-adapted. If we compare the size of the lenses of our small binoculars with our dark-adapted eyes, it's easy to notice that they are 5 times larger in diameter 35 mm ÷ 7 mm = 5). Since we are talking about area, then we must square the difference to come up with 5^2 or 25 times the light-gathering power. For our 10x50's, the lenses are nearly 7.14^2 (50 mm ÷ 7 mm ≈ 7.14) or 51 times greater in light-gathering power. In the same way, a telescope with a 70 mm diameter lens would possess 10^2 or 100x the light-gathering power of our unaided eye. In each succession used in our examples, we have effectively doubled our light-gathering power each time. In astronomy, this means that each larger instrument will enable us to see objects half as bright (or twice as faint) as the former. If you want to see fainter stars and objects and more of them, then you will need a telescope with a larger primary objective.

The second type of telescope uses mirrors rather than lenses to collect and focus light. In a reflecting telescope, the primary mirror has a concave-curved surface, like the working surface of a spoon or dish, so that reflected light is brought to a focus. Similar to the lens of a refractor, the larger the primary mirror the more light-gathering and resolving power it has. The simplest type is known as a Newtonian reflector, named after its inventor Sir Isaac Newton. It uses a primary mirror at the bottom of the tube and a smaller elliptical diagonal mirror near the top to intercept the light path and reflect it out the side of the tube for viewing.

There are other types of reflecting telescopes, most of which belong in a family known as compound telescopes. They are called that because they typically utilize more than one curved reflecting surface. A cassegrain reflector uses a similar primary with a shorter focus than most Newtonians. The short focus is effectively increased dramatically by an outward curved (like the back of a spoon) secondary mirror that intercepts the light path and reflects it back down the tube through a hole in the center of the primary mirror for viewing. Classical, Dall-Kirkham, and other cassegrains are distinguished by the curvature of their secondary mirrors. Another kind of compound telescope is known as a catadioptric, which uses large lenses, or plates, as lens-mirror systems for their optics. The purpose of the specially curved lens is to cancel out the negative effects brought on by the short focus of the primary, known as coma and spherical aberrations. Common examples using lens-mirror designs include Schmidt-cassegrains and Maksutovs. Figure 1 illustrates the optical arrangement of all three basic designs while Figure 2 adds the focused light path.

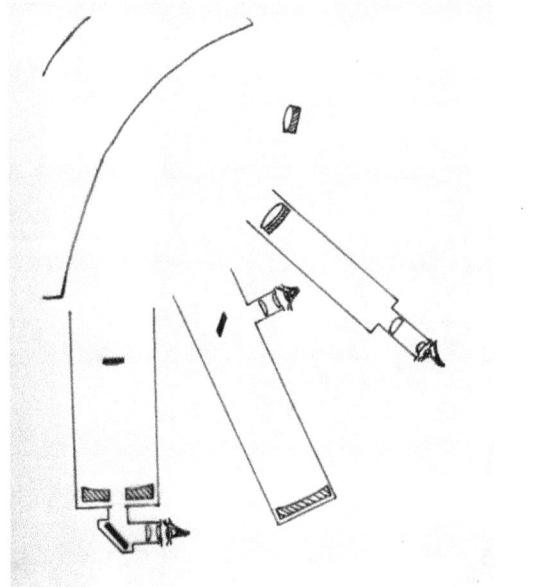

Figure 1

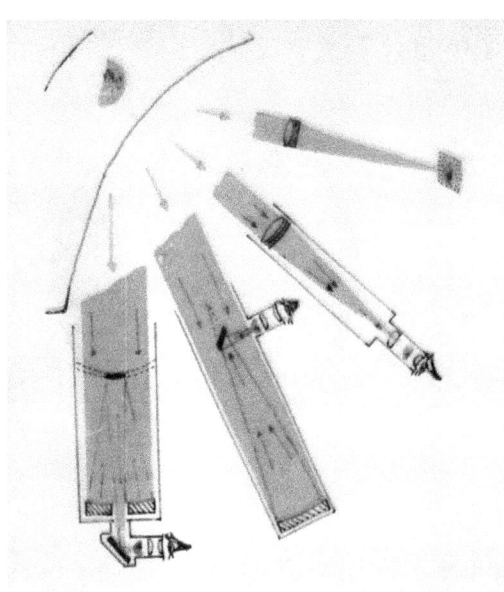

Figure 2

Resolution is the ability to view close-together objects that appear as one into separate distinct objects. Finer details are seen in extended objects, such as planets, as well. Again, the larger the primary objective, the greater the telescope's resolving power. Figures 3-7 illustrates the effects of increased resolution. Figure 3 shows a gray page made from 1 million dots. Figure 4 shows part of the gray page magnified 8 times. Figure 5 shows the same magnified 30 times, and Figure 6 shows it magnified 100 times. If you see a bright star shining steadily in the night sky, you may not realize that the object is actually a planet, such as Jupiter. Point a telescope toward the star, and now it is *resolved* as a disk with features and details with four star-like moons in the same field of view, just as the dots in Figure 6 are now revealed for what they truly are, blotches of ink soaked in paper from an inkjet printer. A photograph of Jupiter in a book shows the same cloud belts and Galilean Satellites, confirming the "star" is really Jupiter. Figure 7 shows a view you may have witnessed in a dark summer sky, a luminous band of light faintly glowing across a large part of the sky. We call it the Milky Way, and it is indeed a section of our galaxy for the glow is made by thousands of *unresolved* stars. Even a small pair of binoculars will *resolve* the glow into separate points of light.

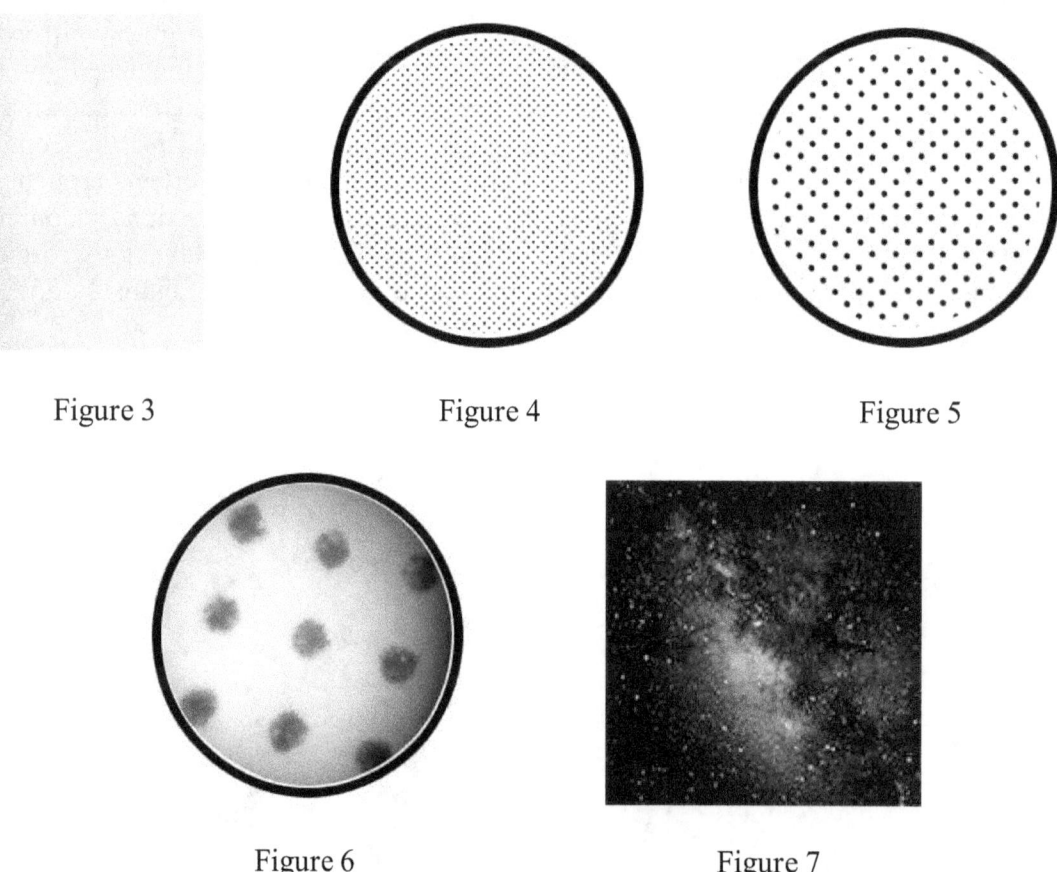

Figure 3   Figure 4   Figure 5

Figure 6   Figure 7

Another general characteristic of any telescope is that it must be able to produce images with good definition. This is determined solely by the quality of the optical system in being able to sharply focus a distinctively clear image of the object in view. Stars appear as pinpoints and the moon and planets present a wealth of detail in subtle features. As formerly mentioned in the introduction, magnification is immaterial when it comes to determining how powerful a telescope is, but is quite necessary in order to enlarge the apparent diameter of an object to an appreciable size for inspection. Within resolution limits, it also helps separate close double or multiple star systems. Finally, another important characteristic of any telescope is that it should preferably have a range, or field of view, wide enough for a number of objects or details to be seen in relation to each other. This is determined in part by the telescope, but also by the type of ocular or eyepiece employed.

Some Useful Definitions & Formulae

Aperture – the diameter of the primary objective

Focal length (fl) – the distance from the primary objective to the focal point

Focal plane – the position where a focused inverted image is formed

Focal point – the point where the converging cone of light rays intercepts and cross each other

Focal ratio – commonly referred to as f-number (f/#), it is the ratio of the focal length divided by the aperture, which is determined by the steepness or depth of curvature of the primary objective
$$f/\# = fl/aperture \textbf{ or } aperture \times f/\# = fl \ (8\text{-inch } f/6 = 48\text{-inch fl})$$

Magnification – the number of times the diameter of an object appears to be enlarged
magnification = fl of telescope ÷ fl of eyepiece
(48-inch fl telescope ÷ ½ inch fl eyepiece = 96X)
(2,000 mm fl telescope ÷ 40 mm fl eyepiece = 50X)

Magnitude – a scale of the relative brightness a star or star-like object appears
Each whole number difference in magnitude corresponds to about 2.512x
The lower the magnitude, the brighter the star, so a 1st magnitude star is about 2.5 times brighter than a 2nd magnitude star, or 6.25 (2.5 x 2.5) times brighter than a 3rd magnitude star (since its logarithmic). A 1st magnitude star is approximately 100 times brighter than a 6th magnitude star, a difference of 5 magnitudes. In about 120 BC, Hipparchus referred to the brightest stars visible to the eye as "first magnitude" and those at the limit of naked-eye visibility as "sixth magnitude". It's like you're on a track team and racing against 5 other runners. When the gun is fired to begin the race, you give it your best and break the tape at the finish line, you're in first place. That means out of all 6 runners, you were the fastest. The sixth place runner who came in last is the slowest runner in the race. Similarly, a 1st magnitude star was considered the brightest star, and a 6th magnitude star the faintest star seen with the unaided eye for the average person on a dark clear night. Today, brighter objects like the planets, moon and sun are assigned magnitudes in the negative numbers, and fainter telescopic stars go well beyond 6th magnitude.

Mounting – used to support, stabilize and point the telescope in any direction from the horizon to the zenith (point directly overhead). The two basic types include **equatorial**, which have a polar axis aligned with the earth's axis for tracking (especially with a clock drive motor synchronous with the earth's rotation) and a declination axis, and an **altazimuth**, which simply moves right-left and up-down.

Optical Tube Assembly – the main part of the telescope housing the optics

Telescope accessories – can enhance observation, such as 1) finder telescopes having a wide field and low power for quickly locating objects, 2) filters that can reduce glare and improve contrast, 3) guidescopes to track objects during astrophotography (taking photographs through a telescope), and 4) star diagonals make it possible to observe zenith objects in comfort for refractors and compound telescopes.

Choosing a Telescope

A telescope is a powerful tool used in exploring objects throughout the universe. They literally enable us to see the invisible! Telescopes are also a time machine, allowing us to peer into the past. The word telescope was derived from the roots tele, which means "distant," and skopos, which means "to see." So a telescope is an instrument that allows us to see distant objects, such as the Moon, planets, stars and star clusters, nebulae and remote galaxies. 2009 marked the 400th anniversary of the first telescope pointed to the night sky by the famous mathematician, scientist and astronomer Galileo Galilei. Modern telescopes are far superior in optical quality than these earlier instruments. Just like the pupil of our eye gets larger in the dark to let in more light, the larger the telescope's optics the more faint light from distant stars and galaxies appear brighter, allowing us to see further and deeper into space. A second benefit with a larger telescope is its ability to resolve smaller and finer details on extended objects like the Moon and planets, and permit the clear separation of close double stars.

Perhaps you are interested in purchasing a new telescope. Basically, there are two types of telescopes to choose from. A refractor uses lenses to collect and bend light as a cone to a focus. Binoculars are merely two refractor telescopes mounted side by side. Reflectors use a set of mirrors to gather light, which is brought to a focus by virtue of a concave curve (inward like the scoop of a spoon) on the front surface of the primary (largest) mirror. Light enters a mostly hollow tube and reaches the primary mirror at the bottom. As the reflected cone of light (due to the curve) travels up the tube, it is intercepted by a smaller flat (plane) diagonal mirror set at a 45 degree angle with respect to the light path. So, 45 + 45 = 90 degrees and the light path is sent outside the tube at a right angle for the observer to inspect a focused image through an eyepiece (ocular). This is a classic Newtonian reflector, named after another famous scientist, Isaac Newton who created its design. The distance between the primary objective (lens or mirror) and the eyepiece where the focal point is reached is called focal length. This is determined by how steep or shallow the curve in the glass is. A greater curve will focus light in a short distance, so the telescope tube will be correspondingly shorter as well. A shallow curve will extend this distance, calling for a longer tube assembly. Many reflectors are referred to as compound telescopes because of their short, stubby tubes. This cassegrain design uses a steeply curved primary mirror and a convex (curved outward like a ball) secondary mirror mounted near the top center of the tube. When light reflects off of this convex curved mirror, the steeply converging (come together) rays of light are made to diverge (spread apart), thereby effectively extending the focus further so the light path will continue through a central hole in the primary mirror (like a donut) and focus outside the rear of the tube assembly. Many cassegrains use a special glass plate at the front of the tube to "correct" the light path from different problems inherent in this design. They may be called a Schmidt Cassegrain or a Maksutov.

In order to point the telescope's optical tube assembly at a particular location in the night sky, it will need a mounting. There are basically two types of telescope mountings. An altazimuth mounting has two axes at right angles to each other where one axis allows the telescope to pivot up and down (altitude) and the other axis left and right (azimuth). It's the simpler of the two. The other type is called an equatorial mounting. It also uses two axes at right angles to each other, but one of them, called the polar axis, is set in line with the earth's axis of rotation. Once accomplished, you simply set the declination (north-south) axis and right ascension (east-west) on the polar axis to point at a particular object, then just rotate westward on the polar axis to track an object in the sky as it appears to move due to the earth's rotation. Setting circles may be attached to both axes for locating objects using their celestial coordinates (right ascension and declination). If the polar axis has a clock drive motor, it will automatically guide this tracking at the same rate the earth is turning. Many commercially-made telescopes now come with computer controlled guiding systems and a push-button hand paddle known as "Go-to" capability. This is great for taking pictures through the telescope, known as astrophotography. If not, hand knobs with worm and gears are typically used to manually guide the instrument. Either way, the mounting is supported typically on a pedestal or tripod. Some common types of equatorial mountings include the German, fork, English yoke, and others.

Many accessories are available or required to properly operate a telescope. A finder can either be a small refractor telescope with a wide field of view and crosshairs or a laser device used to point the main telescope accurately and "find" the object sought for viewing. Eyepieces come in different types and sizes. Magnification is calculated by dividing the focal length of the telescope by the focal length of the eyepiece. For example, a telescope with focal length 900mm using an eyepiece with a focal length of 20mm will yield a magnification of 45X (900/20=45). Another way to say this is 45 power, which means that objects will appear 45 times larger in diameter than with the unaided eye. A common misconception is that magnification determines how powerful a telescope is. Since magnification can be adjusted using different focal length eyepieces for any telescope, the telescope's true "power" is determined by its size (aperture or diameter). A Barlow lens can amplify the magnification of any given eyepiece by effectively extending the focal length of the telescope. Commonly, they double or triple the magnification of the eyepiece used, so in our previous example we now have 90X or 135X with the same eyepiece. Filters are typically threaded to screw into the barrels of eyepieces and come in different colors to enhance specific features of planetary detail. Lunar filters work well to reduce glare and increase contrast on the Moon. Solar filters block all the harmful rays allowing safe viewing of sunspots. Other filters are used for reducing scattered light as in city areas permitting views of faint extended objects such as nebulae. A star diagonal prism or mirror changes the position of the eyepiece by creating a right angle to the normal light path. This works well with refractors or cassegrain reflectors, especially when viewing objects high overhead. Dew caps extend the tubes length preventing dew from forming on the surface of lenses or corrector plates.

What advantages/disadvantages exist for either type telescope design? Well, if you want as large and powerful a telescope as feasible within your budget, you must consider whether you intend to place it in the car to transport it to a dark sky site or you intend to permanently mount it in your back yard or observatory. Again, a long-focus telescope will have a longer tube, possibly prohibiting it from fitting in the car easily. A short-focus telescope, sometimes called richest field telescopes or RFT's, are great for looking at wider fields of view in the sky, capturing beautiful views of rich star clusters and several objects in relation to each other simultaneously. Long-focus telescopes are usually superior for examining fine details on planets and splitting close double stars. The trade-off is loss of portability. Refractors have an optical advantage over reflectors in that they don't suffer diffraction (light scattered or bent around the edge of a barrier) caused by the central obstruction to the incoming light by the diagonal or secondary mirror. If the lens is of high optical quality, the refractor will typically outperform an equal sized reflector in producing sharply-defined images. Refractors are commonly made with a long focus, making the tube length a concern. Also, refractors are far more expensive than reflectors of equal size. Since the size of the telescope's lens or mirror is a function of its light-gathering power, the reflector is preferred overall when larger sizes are desired.

How about some tips for enhancing your observational experience through a telescope? When beginning any observing session, always start with the longest focal length eyepiece and the widest field of view to help spot objects more easily in the telescope. Simply align the telescope by sighting along the top of the tube and pointing it in the general direction of the object in the sky. If you have an equatorial mounting with setting circles, you can look up the celestial coordinates of the object and adjust accordingly. Of course, if you have Go to capability and have done a 2 or 3 star alignment, you may simply push a button on the hand paddle. Either way, your next step is to spot the object in the finder and align so it appears in the center of the crosshairs (like a gun sight). If your finder is optically aligned with the main telescope, the object should appear in the field of view of the eyepiece. When observing faint objects try using a technique known as averted vision. Instead of looking directly at an object, look off to the side a bit and see if you notice that the object appears brighter. That's because that part of your eye's retina has more cones, which are sensitive to light and dark. Become familiar with the night sky by using a planisphere, commonly referred to as a star finder. Get to know the brightest stars and seasonal constellations by name. Attend a planetarium show to learn their relative positions in the sky. It takes time, but the universe is a very patient place, one that doesn't mind waiting while we take the first steps towards understanding.

Using a Telescope to View the Night Sky

On a clear, moonless night, the average person can see between 3 and 4 thousand stars across the sky. It's as if we're inside of some great celestial (sky) sphere upon which the stars appear to be attached. Of course, stars are actually at different distances from us in a 3-dimensional understanding. In science we like to use models to illustrate abstract concepts. A model of the celestial sphere has a ball representing the earth centered inside of a larger ball with stars painted in patterns we call constellations. The International Astronomical Union officially designated a total of 88 constellations in the entire celestial sphere in 1930. If we took our model and crushed it into a flat, two dimensional plane, then we would have created a planisphere. A circular cardboard cutout looking like a star map freely rotating on a central pivot and connected to a square cardboard piece can be set to a time and day of the month while facing a certain direction. This device helps in identifying bright stars and constellations in the night sky, and is commonly called a starfinder.

Recognizing stars and constellations is a good start, but locating much smaller, specific areas of the sky requires a detailed set of star charts. Just like there are maps helping us find our way in unknown geographic locations on the surface of the earth, we could create maps that indicate relative positions of objects in the night sky as well. Lines representing streets and roads going north-south and east-west on our earth map can be turned into lines representing celestial coordinates on our sky map. The fictitious lines of latitude and longitude crisscrossing our earth map can be thrown up into the sky directly overhead. These celestial lines of latitude north and south of the equator are now called lines of declination. They measure north of the celestial equator from 0 degrees to +90 degrees for the north celestial pole and south from 0 degrees to -90 degrees for the south celestial pole. Each degree can be further subdivided into 60 equal minutes (not time) of arc (part of the circle formed by the complete line) and each minute divided into 60 equal seconds (again, not time) of arc. The Earth rotates on its polar axis through fifteen minutes of arc in every minute of sidereal time (the motion of the stars due to the rotation of the earth). One minute of arc at the Earth's celestial equator is approximately equal to one nautical mile on the earth. The celestial lines of longitude east and west of the prime meridian, a line representing 0 hours or the beginning of a new day over Greenwich, England, are now called right ascension. They measure hour angles increasing westward from 1 to 24 (0 and 24 are the same line) and are subdivided further into minutes and seconds of actual time.

Using this celestial coordinate system along with maps called star charts; we can pinpoint the location in the sky of any celestial object of interest. Equatorial mountings for telescopes can incorporate setting circles to manually align the instrument or use computers and motors to locate and track celestial objects. Sextants and astrolabes are used in celestial navigation to compare a star's known declination with its angle above the north-south horizon.

Now let's get started with listing some materials and equipment we will need to construct our very own telescope in chapter 2.

Let's Get Started

This book serves as a guide in building a Newtonian reflector, with an 8-inch as the example. In order to successfully complete building your own telescope, there are a number of parts and tools you will need to gather together. The first determination you need to make is whether you aspire to grinding your own primary mirror. Chapters 6-8 cover the details of constructing the primary. Grinding your own mirror is a tedious, painstaking, and lengthy process, but well worth the effort for the satisfaction of knowing you made the heart of your own telescope. The choice is yours to make, and even if you choose to purchase a finished mirror from a reputable optical company (as with everything else, I will give you a good list of sources), you will still enjoy the deep satisfaction of building the telescope itself.

Parts
Sonotube
Mirror cell
Diagonal mirror
Diagonal mirror holder and spider
Focuser
Mirror or mirror making kit
Finder telescope
4x8 foot sheet of ¾-inch oak plywood
Teflon pads
Ebony Star Formica
Tightbond wood glue
Philips Drywall Screws #6
Chest handles
Course, medium and fine sandpaper
Sanding Sealer
Polyurethane
Contact cement
Flat black paint
Gloss white paint
White primer
Machine hex screws Center
bolt and tee nut Eyepieces
& Barlow lenses Circular
protractor

Tools
Table saw
Saber (jig) saw or Router
Radial arm or circular saw
Drill press or handheld drill
Paintbrushes
3-inch heavy nap roller with extension handle

The use of each of the tools and materials will be described in great detail as we go step by step in constructing a beautiful instrument designed to take us on a journey through the universe. Chapter 3 will provide an overview laying out the plans for this wonderful project.

Unravel the Plans

Since the entire telescope mounting and some of the tube is made of wood, we'll begin by deciding how to make the most of our 4x8 foot sheet of ¾-inch oak plywood so all the pieces can be cut from that one sheet. A good start is to use graph paper or a clean sheet of paper with a ruler to measure the dimensions of all pieces making up each part. All measurements are done in English feet and inches. So let's list the parts with the measurements of each piece, and letters matching them to the illustration below:

Ground Board
1 square 15 ½ in. x 15 in. (a)
4 feet each 2 ⅞ in. x 2 ⅞ in. (b)

Rocker Box
1 bottom 15 ½ in. x 15 in. (c)
2 sides each 15 in. x 26 in. (d)
1 front brace 19 in. x 14 in. (e)
1 rear brace 14 in. x 2 ⅝ in. (f)

Tube Parts
2 end rings each 13 ⅛ in. x 13 ⅛ in. x 1 7/16 in. wide (g)
1 mirror cell support 13 ⅛ in. x 13 ⅛ in. (h)
2 trunnions each with 4 pieces:
　　► bracket 12 in. x 6 in. (i)
　　► 2 end contacts 1 ⅞ in. (⅞ in. mid) x 6 in. (j)
　　► circle 7 in. x 7 in. (k)

Each of these 3 parts shall be illustrated in detail later as we focus on them one at a time. Our purpose now is only to utilize every square inch of our sheet of plywood. So let's see what our drawing looks like so the sheet of plywood can be cut correctly in Figure 8.

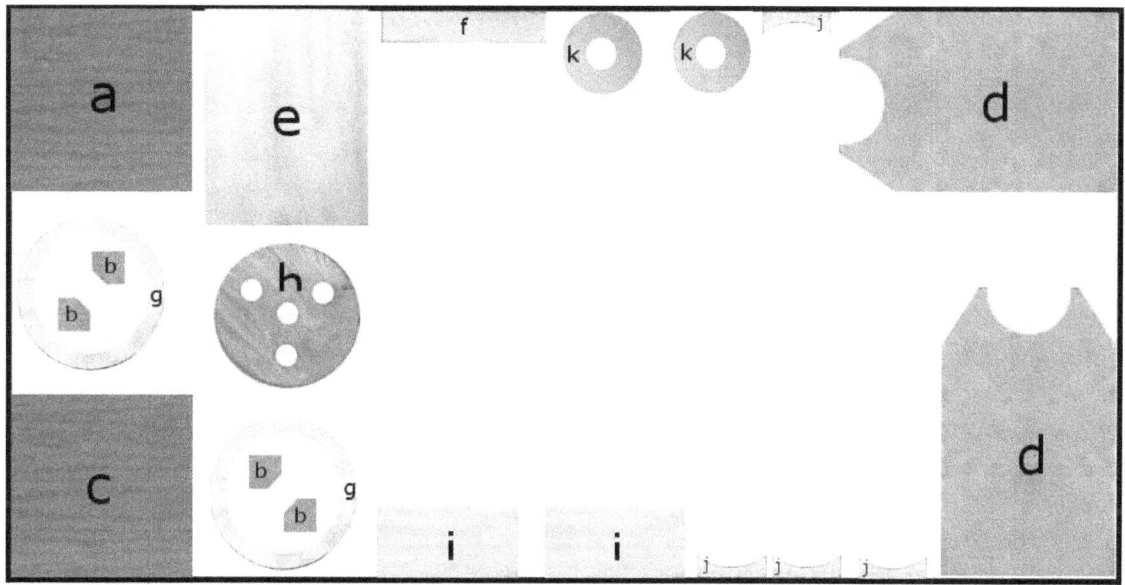

Figure 8

As you can see, there is quite a bit of the plywood sheet left. All preliminary cuts will be square, so a table saw (preferred) or circular saw will be used as our first tool. A radial arm saw can be used later for smaller pieces if you have access to one. The pieces are arranged so corners and straight edges are taken advantaged of and the number of precise cuts and amount of sanding later is minimized.

Hardwood plywood is a little more expensive than regular plywood, but is usually very flat and smooth with no knots, and has superior strength for long-lasting quality. Oak or mahogany plywood is a good choice for its beautiful wood grain appearance, strength and durability. Other tools that may come in handy include a tape measure, yardstick, pencil, square, and a straight edge such as a bubble level. See Figure 9 below.

Figure 9

A full sheet of oak plywood is a bit heavy, and a few extra hands will make it easier when first cutting into it. A large and stable table saw with a good sharp blade is preferred for these first cuts. If you are using a circular saw then at least a couple of saw horses will help. Take things slowly and always keep safety at the forefront of your mind. You cannot afford to be careless around any kind of saw at any time. Perhaps you can turn this into a school project and ask if you can use the school's wood shop, which are typically well-equipped. The first cut can be made to divide the 4x8 foot sheet into two 4x4 foot sheets by cutting it in half. It's much easier to move around when cutting out the smaller pieces. One of the 4x4 foot halves can be cut in half again for two 2x4 foot sections. They will prove even easier to manage when making smaller cuts. Figure 10 on the next page illustrates the manhandling of the full-sized sheet of oak plywood during its first encounter with a saw blade.

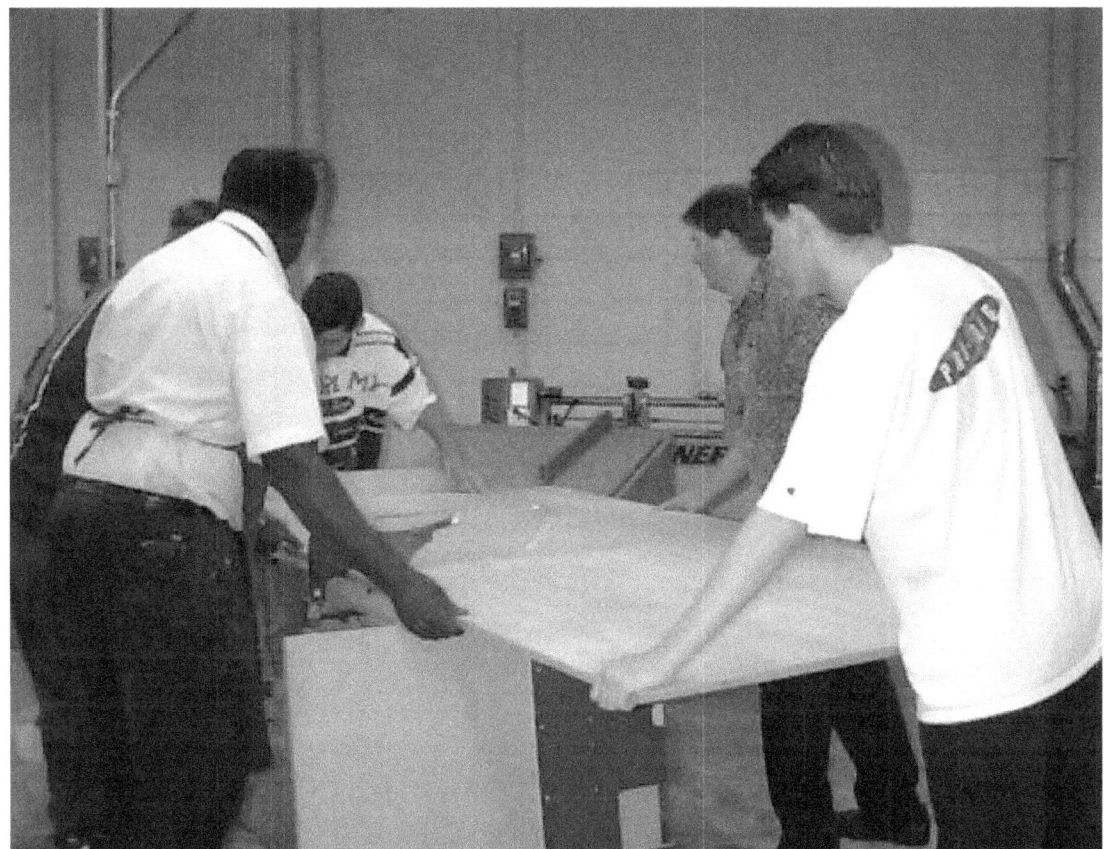

Figure 10

Other than making nice straight cuts in the beginning, get the general pieces cut out first for more precise shaping of the wood later when they are more easily manageable. See Figures 11 and 12 below.

Figure 11

Figure 12

Now we turn our attention to shaping each piece of wood into the sections that will be sanded and assembled together using wood glue and screws into the main parts of the completed telescope. We will start from the ground and work our way up in chapter 4.

From the Ground Board Up!

The Ground Board is the very base that supports the entire telescope on a fairly level surface of the ground or pavement using four small "feet" that functions as a rigid table supporting the weight and pivots of the telescope above. It's essentially one simple square cut of plywood with four smaller square pieces mounted in each corner. Not exactly a perfect square, the 15 in. x 15 ½ in. dimensions of the ground board are such in order to match the size of the part that rides above. It is pictured in Figure 13 below.

Figure 13

Four small square cutouts just under 3 inches are mounted to each of the four corners of the main board to act as feet interfacing with the ground itself. To help lessen the weight you can cut off one of the corners diagonally from each foot and face it inwards towards the center of the ground board. See Figure 14 below.

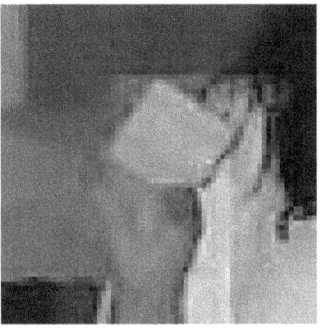

Figure 14

Figure 15 shows the completed wood part of the ground board from the side facing down or toward the ground. Wood glue and screws were used to fasten the feet into place. Before doing this with any of the parts, you want to sand the surfaces of each of the wood pieces using fine sandpaper on the faces and medium sandpaper on the edges. Once smooth, blow the sawdust off using an air compressor or simply wipe off with a clean rag.

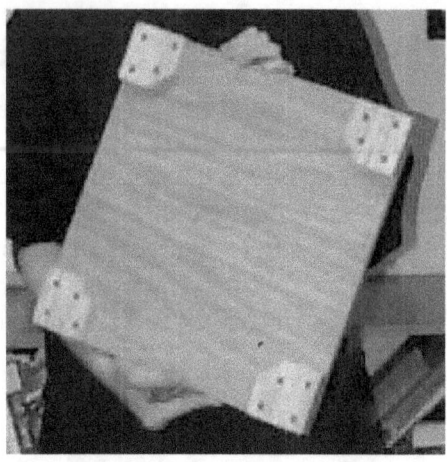

Figure 15

Mark with a pencil the positions for all screws, and drill pilot holes using a drill bit slightly thicker than the shaft of the screw not counting the threads. This will prevent the wood from splitting when turning the screws down tighter. Coat both surfaces with wood glue and use Philips Drywall Screws - #6 x 1 1/4" are popular for use in cabinet and speaker enclosure assembly with a black oxide finish and can typically be purchased 100 per box. See Figures 16 & 17 below. If you don't already have one, I highly recommend you get a philips head drill bit so you can get the job done quickly and save your wrists from doing it by hand. Let the drill do the work. Once assembled and the glue has dried, paint a coating of Sanding Sealer over all surfaces. Let dry, sand with fine sandpaper, and brush a final coat of Polyurethane for a glossy wood-grain finish.

Figure 16   Figure 17

The first ground board prototype I put together is shown in the upper left of Figure 18. A circular plywood base was mounted to flanges and steel pipe to extend the legs outward in three directions, like a tripod. Through trial and error, I found it was not very sturdy as needed for the base support of any magnifying telescope. The second ground board was of standard design as a square base with four feet in the corners, as pictured in the center of Figure 18 below. This was for a large heavy telescope so I doubled the layers of plywood for the base and the feet, which may have been unnecessary and only resulted in doubling the weight. This ground board is quite heavy, but I know you could practically put a house on top of it and it wouldn't budge. The ground board for our much smaller 8-inch telescope will certainly not need more than one layer of plywood all around.

Figure 18

Figure 19 shows the underside of our 8-inch finished ground board and as you can see, the four feet have already taken a beating on top of all kinds of surfaces to set up the telescope. Well, it should hopefully last a lifetime!

Figure 19

Figure 20 shows the topside of this ground board. The three white square objects mounted onto the top surface are equidistant apart in separate 120 degree lines along the perimeter of the board. These pads are made of pure virgin Teflon and will interface with the underside of what sits on top of the ground board, the Rocker Box, which has rough-textured Formica countertop contact cemented to its bottom. The coefficient of friction between these two surfaces makes for a perfect bearing that slides as the rocker box turns left and right in azimuth.

Figure 20

The Teflon pads are screwed and contact cemented into place and the hole drilled through the center of the ground board platform is to secure the rocker box on top. A tee nut is hammered into this hole from the bottom or ground-facing side so that a bolt can screw into its threads from the topside. The bolt should not be tightened too much because this will hold the rocker box to the ground board but we want to allow it to swivel over the ground board's center without hindrance. Mount one of the three square Teflon pads using contact cement and a short screw (⅝") through the central hole in the middle of the board facing one of its sides. The distance from the middle of the central hole to the screw hole in this square pad is exactly 5 ¾ inches. The same will be the case for the other two square Teflon pads using a circular protractor to arrange those 120° apart. This will allow the bottom of the rocker box to always be in contact with all three pads simultaneously, which is critical for stability. Teflon pads and center bolts with tee nuts are typically sold as a kit by some telescope making companies, such as AstroSystems and Taychert Telescopes. AstroSystems sells a Pivot Kit with a ⅜" x 2.5" brass bolt, bronze thrust bushing, stainless steel locknut assembly and full instructions. Though the bushing and Teflon pad are not needed under the center bolt, it does improve telescope motion. At the time of this writing it sells for $27.00. They also sell Teflon Bearing Kits, in which case the Small kit is all that is needed here. The kit currently sells for $22.00 and contains 3 azimuth and 4 altitude bearing pads of virgin Teflon, pre-drilled and countersunk, with stainless steel mounting screws. These can be found online at the web site for AstroSystems at http://www.astrosystems.biz/pivot.htm. Taychert Telescopes also sells an azimuth bearing and pivot kit for mid-sized Dobsonian telescopes with instructions for $13.50 + shipping and can also be found online at http://www.tscopes.com/dobparts.html#teflonkits. ScopeStuff offers four virgin white PTFE Pads, 1/8" Thick, 3/4" x 1", drilled and countersunk, with screws for $17.00 at http://www.scopestuff.com/ss_lexx.htm. These work the same as the Teflon pads. Figure 21 shows a close-up of the hole for the center bolt.

Figure 21

Figure 22 below shows a close-up of the tee nut on the bottom of the ground board. Notice the two Phillips head wood screws holding it in position.

Figure 22

This completes our discussion of the ground board, which has the job of sturdily supporting all the combined weight and motions of the telescope above it by providing a rock-solid platform and smooth bearing surface for the entire telescope to easily pivot on.

Now we will turn our attention to the part that glides over the surface of the ground board, the Rocker Box assembly, next in chapter 5.

Rocker Box & Roll

The Rocker Box serves two primary functions acting as a "middle man" between the ground board stand and the telescope itself. It allows the telescope to pivot right and left in azimuth and up and down in altitude. That's why this type of mounting is categorized as an altazimuth. This simple yet very sturdy design mounting was created by John Dobson, and in his honor is called a Dobsonian. This telescope mounting design has literally revolutionized the advent of exploding numbers of amateur telescope makers (ATM) around the world. Using some modern day gadgets like stepper motors and digital setting circles, the Dobsonian altazimuth mounting can be transformed into an instrument that can locate, track and even photograph (using a field derotator) celestial objects with ease. But the scope of this book is to keep things simple so anyone can build and enjoy their own powerful telescope that will outperform by far the off-the-shelf telescopes found at your local department store. You will also find it much more user-friendly and easy to set up in moments, making for a fulfilling and thoroughly enjoyable experience for all.

The height of my rocker box was designed for an 8-inch f/8 Newtonian reflector telescope. Isaac Newton created this design many years ago. The f-number is called the focal ratio of the telescope and is determined by how steep or shallow the curvature of the primary mirror is. The front surface of the mirror has this concave curve (like the inside of a spoon) so as to bring the light to a focus. Steep curves have a low f-number while shallower curves have higher f-numbers. The focal ratio also determines the focal length of the telescope. To calculate this distance, simply multiply the aperture or diameter of the mirror by the focal ratio. In my case, 8 inches times f/8 = 64 inches focal length (fl). See Figure 23 below. This approximates the length of tube required to house the optical system, and is called an optical tube assembly (OTA). Since I am tall and the eyepiece of a Newtonian telescope is found near the top of the tube facing upwards toward the sky, the f/8 system works well for me. However, the real reason I chose to make my telescope an f/8 as opposed to say, an f/5 or 6, is because the longer focal length telescopes on average outperform their shorter counterparts in revealing subtle detail on planets, given the same optical quality. The shorter or faster (term used in photography) systems usually display brighter images over a wider field of view, making it the preference for viewing rich star clusters and extended nebulae. So it's a matter of choice of what kinds of viewing you intend to spend most of your time on and how long, or high, you want the tube and eyepiece to be. Many people choose an 8-inch f/6 telescope, making the focal length reduced to 48 inches, or a tube length just over 4 feet. If you decide this is what's best for you, then you can reduce the height of your rocker box accordingly. For now, we will stick to the height of mine, which is two feet two inches above the floor of the rocker box.

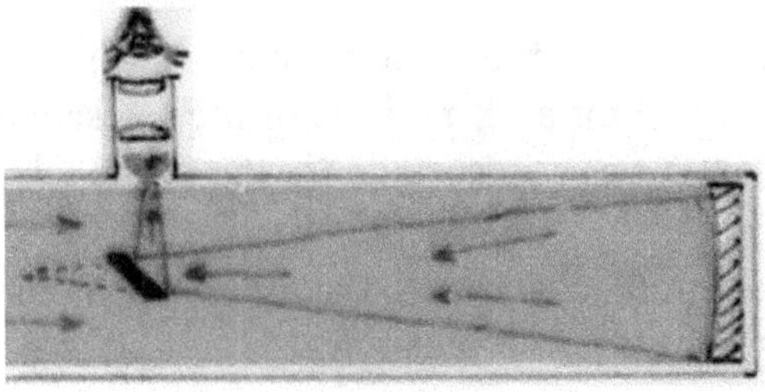

Figure 23

Now, let's begin by taking our two matching sides and shaping them for the altitude bearings the telescope will ride up and down on. Right now, they are simply a pair of rectangular pieces of wood with the dimensions of 15 in. x 26 in. and one end of their lengths must be cut a certain way. The first thing we must do is decide how wide of a semicircular cut we must make so that the trunnions of the OTA will fit inside. Those trunnions are going to be exactly 7 inches in diameter. When we add the thickness of the Formica laminate (1/16") and the Teflon pads (1/8"), then we see that our semicircular cuts must equal that of a circle 7 ⅜" in diameter. A router would be best for making this cut, but I used a Porter-Cable jigsaw with a bracket and pin to make the circular cut required. Also, to lighten the load once again and simultaneously make it look less boxy, I trimmed the corners toward the tops of the semicircular cut on each. The result is what you see below in Figures 24 and 25.

Figure 24

Figure 25

When the raw cutting is done, then we want to sand the pieces smooth with no jagged edges and then blow or wipe off the sawdust (Figure 26) so we have a finished product ready for assembly (Figure 27).

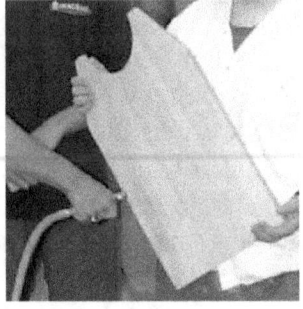
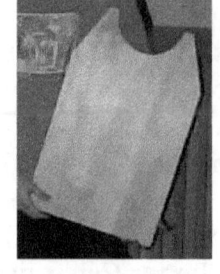

Figure 26   Figure 27

Next, let's assemble the other three pieces of the rocker box together after sanding them using wood glue and screws. Remember to always pencil mark and drill holes where the screws will go **before** putting the glue on each surface. Figure 28 shows the bottom floor, front and rear braces assembled for drying. Tightening the screws compresses the two adjoining pieces coated with glue togther, forcing the excess glue out the sides requiring a wet rag for cleanup, and should result in the pieces being at right angles to each other. A level can always be used to check this.

Figure 28

Before the glue has a chance to dry, add the two sides with glue and screws to complete the rocker box assembly, as shown in Figure 29 below.

Figure 29

Figure 30 shows the front view of the assembled rocker box. Notice that the front brace allows for enough clearance so that the OTA can point all the way down to the horizon, while the rear brace allows the bottom of the OTA to clear without interference.

Figure 30

Figure 31 shows the rocker box after sanding sealer and polyurethane have been appiled for a nice finish. The bottom of the floor, however, has not been coated so that a layer of Ebony Star Formica laminate can be attached to the wood surface using contact cement, as shown in Figure 32.

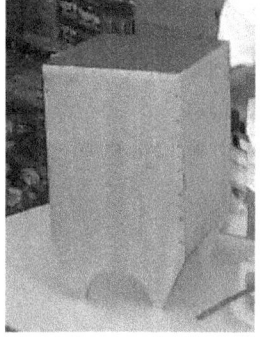

Figure 31   Figure 32

The best tools for cutting the laminate to size include a utility knife and an aluminum yardstick or other straight edge. Always cut from the unfinished backside of the laminate, not the crinkled and colored finished side. Always exercise safety when using a utility knife as the razor blade is very sharp. It helps to have the laminate face down on an old piece of carpet while cutting, and to begin gently scoring a line a few times rather than applying a lot of pressure to cut through the first attempt. Wilson Art brand Ebony Star style kitchen counter laminate 4552-50 doesn't seem to be as widely available as it used to be. Possessing a slight texture and "bumpiness" seems to be an important element to get the correct feel in the finished telescope. Teflon is extremely slippery and does not stick to other materials, so the Formica sets the tone with the right amount of friction with no backlash. Smoother brands of this are available almost everywhere building supplies are sold, but they won't do as well as the old stuff. A company known as ScopeStuff sells the right kind of Ebony Star as disks rather than sheets, though they may sell you a sheet of it if you inquire. ScopeStuff can be reached at http://www.scopestuff.com/ss_lexx.htm on the web. A photo of different sized disks is shown in Figure 33 below. The smallest one is sufficient for this project at 13 ½ inches outside diameter.

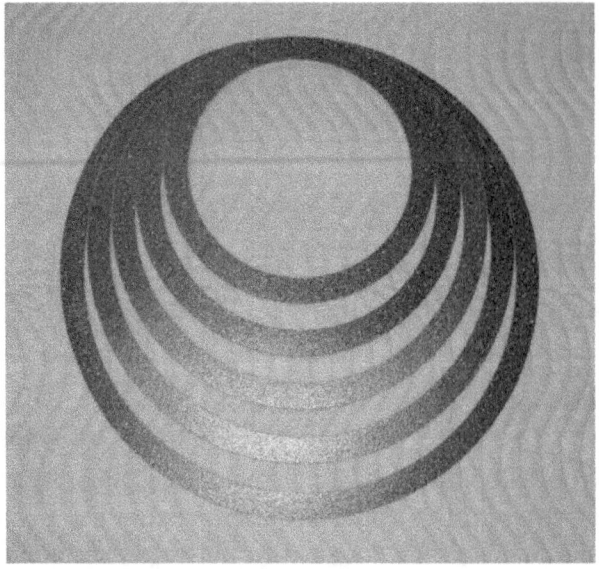

Figure 33

An alternative could be Glassboard FRP as your choice of bearing material, though it is usually considered a great substitute for large aperture telescopes. Either way, after the contact cement has set, we will need to drill a hole through the center where the center bolt will connect it to the ground board. See Figure 34 below.

Figure 34

Steel chest handles can now be screwed into each side of the rocker box for easy carrying, as can be seen in Figures 35 and 36 following. Please disregard my highly-weathered lawn while living out in the arid mid-west for a period of time. ☺

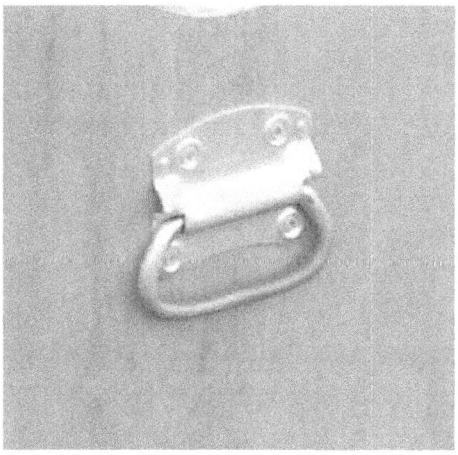

Figures 35

Figures 36

Also notice in Figure 37 the center bolt connecting the rocker box to the ground board.

Figure 37

Now we are ready to cement and screw into place the pure virgin Teflon pads within the semicircular cuts to act as the altitude bearing surface. If you want, you can notch out a little wood to sink them into position so they will never shift, though the contact cement will most likely hold them in place permanently. A general rule of thumb on positioning these pads is that the higher up you place them the greater the tension, and consequently less slippage of the telescopes balance at various altitudes. Figure 38 shows the Teflon pads in position.

Figure 38

Figure 39 shows us the completed rocker box connected to the ground board via the center bolt turned hand-tight into the tee nut, thereby holding the two together as a unit and yet free to swivel right and left with ease and fluidity of motion. Figure 40 shows the inside view of the finished rocker box ready for action. For the next three chapters, we will discuss the "eye" of the telescope itself.

Figure 39

Figure 40

Grinding Away

The heart of any telescope is its optics. Indeed, every other part has the sole purpose of holding the optics in alignment and pointing them in the precise direction of various objects all across the night sky. So before we discuss the entire optical tube assembly, let's look into constructing the primary mirror of our reflecting telescope. The limits of this book to keep things simple prohibits an exhaustive and detailed description of all the technical aspects of telescope optics, and perhaps most of us would opt out of this tedious task in lieu of purchasing a finished mirror from a reputable optical company, in which case you would not wish to know and understand every aspect, so I will condense the next three chapters to cover the basics only.

Two principle magazines that carry telescope making information that I have used over the years include Sky and Telescope's "Gleanings for ATM's" from November 1941 through June 1998, and the complete run of Telescope Making magazine. Books devoted to a complete and thorough description of how to grind and polish a parabolic primary mirror are available. Two of them I personally own have a reputation for setting the foundation on this topic. The first edition of *How to Make a Telescope* by Jean Texereau has been universally accepted as the best book ever written for making a Newtonian reflector telescope. The newer 2nd Edition that I own is much larger and includes new chapters on making different kinds of telescopes and telescope mountings, numerous accessories and modern techniques, and using the telescope. It is considered the most complete single work available on making reflecting telescopes. Both the cover I own and an updated cover for the 2nd Edition are pictured below in Figures 41 and 42.

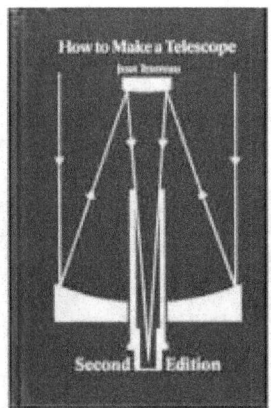
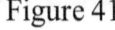

Figure 41

Figure 42

Another book I own and recommend is *Build Your Own Telescope* by Richard Berry. The author uses clear, step-by-step instructions and numerous photographs to describe how to build five different telescopes using ordinary household tools and materials. The user-friendly style helps keep you engaged in the process. Again, both the older original cover of the one I own and the newer modern cover are pictured in Figures 43 and 44.

Figure 43   Figure 44

Willmann-Bell, Inc. has been around for years and have great telescope making supplies, and for the last 36 years have offered the finest in complete Mirror Kits. I purchased the 8-inch Willmann-Bell Pyrex® Mirror Kit years ago to make the mirror for the telescope I am writing about in this book. One of the things that impressed me about this kit is that both the mirror blank and tool were made of top-quality fine annealed full-thickness Pyrex® glass substrate. Before this, the tool I received with such kits was thin and only made of plate glass. Pyrex® is a low-expansion glass with excellent optical properties and a common choice for making high-quality telescope mirrors. You can purchase one of these 8-inch mirror kits yourself from Willmann-Bell, Inc. currently for $145.95 online by visiting http://www.willbell.com/ATMSupplies/ATM_Supplies.htm. The entire kit is pictured below in Figure 45.

Figure 45

The purpose here is to set some abrasive material between two glass surfaces so that when one is moved over the surface of the other, a curve is formed. The top glass blank (mirror) that is face down and being moved across the top surface of the other stationary glass blank (tool) has its center in contact with the edges of the other more often, resulting in a concave (curved inward like the business end of a spoon) curve for the mirror and an equally convex (curved outward like a ball) curve for the tool. Each kit pictured above includes the following:

- Two fine annealed Pyrex blanks with 4 workable surfaces
- Fast-cutting, long-lasting silicon carbide
- White aluminum oxide lapping powders
- Fast polishing cerium oxide
- Micro-facet netting for the pitch lap
- Tempered burgundy pitch for polishing
- Pure packaging to prevent contamination

Since the first order of business is to calculate the depth of the center of curvature (Sagitta) so we know what we are shooting for, and then to grind that curve in the glass, the largest sized grit abrasive is used in the beginning so we can cut as much glass away as quickly as possible. This stage is known typically as "rough grinding." See Figure 46 below for a sample of this curve, though much more accentuated then the curve of our mirror. As was mentioned in the previous chapter, the f-number or focal ratio determines the focal length (fl) of the telescope. Think of a round glass or plastic bowl with a lid on it, like Tupperware. The Sagitta is the distance or displacement from the bottom middle of the bowl to the top lid. Once setting the curve, successively finer grades of grit (smaller sized abrasive particles) are used to reduce or eliminate pits and scratches in the glass cut by the former larger grit. To calculate the Sagitta of an arc, we use the formula:

$$s = r - \sqrt{r^2 - l^2}$$

Where (see Figure 47 below):
s = the Sagitta (sag) or displacement
r = the radius of the arc (the focal length we desire)
l = ½ the length of the chord (span) connecting the two ends of the arc (½ diameter of mirror = 4 inches)

Figure 46

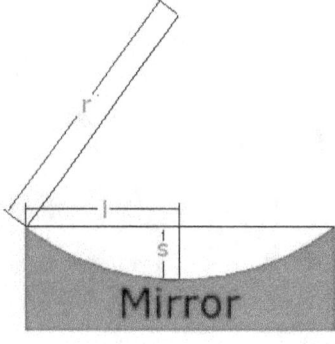

Figure 47

Using our formula for Sagitta, if we want an f/6 mirror with a focal length (r) of 48", then our Sagitta = 0.16695702759441", or rounded to 0.167". If we would prefer an f/7 system with an "r" of 56", then Sagitta = 0.14303982492424" or 0.143" rounded. For my f/8 telescope, Sagitta = 0.12512230931475" or 0.125" rounded, which is ⅛". A practical way to measure this distance is by combining the blades of a feeler gauge to equal the Sagitta and gently sliding them under a straight edge set across the middle of the mirror. Once it fits and that depth is reached, you're done with rough grinding.

What is needed to grind and polish your own reflecting telescope optics, the primary mirror in particular, is some materials and plenty of workspace to permit the freedom to walk around a central area without restriction. A steel barrel as seen in Figure 48 below serves as a perfect platform to work around. Notice a plywood disk was cut to fit tightly inside the top rim of the barrel, and three wood cleats are screwed into position loosely around the perimeter of the tool to act as braces. A wooden wedge will hold the tool blank still when tapped between the mirror side and one of the cleats. See the close-up in Figure 49.

Figure 48

Figure 49

So let's place the tool into position on our new working surface and wedge tightly. See Figure 50 below.

Figure 50

Before we can begin grinding the mirror blank over the tool, we must bevel the edge of the side that will get ground to a curve. This is simply to prevent chipping of the mirror's edge. To accomplish this, we will use a carborundum or silicon carbide (basically sand) stone. Figures 51 and 52 shows how it is done.

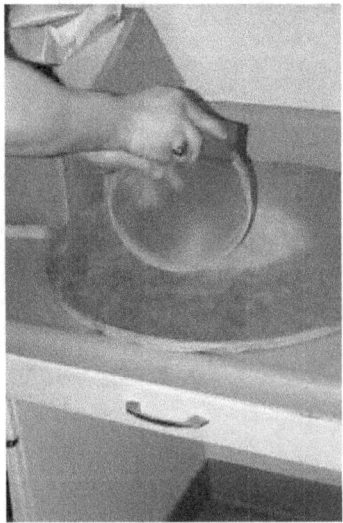
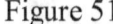
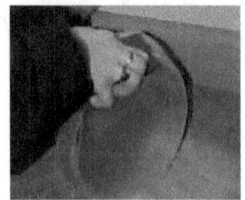

Figure 51   Figure 52

Now we are ready to begin grinding the mirror using the coarsest grit with the largest abrasive particles. Willmann-Bell seals their abrasives in a heavy-duty plastic bag to ensure its purity. Professional mirror makers and opticians transfer the abrasive to a small bowl and apply it to the surface of the tool by spoon or make it a slurry and apply it by brush. Using a spoon, scoop some of the abrasive (about half a teaspoon) and sprinkle the particles over the surface of the tool and then using a trigger sprayer bottle containing distilled water spray some of the water over the entire surface of the tool onto the abrasive, as shown in Figure 53 here.

Figure 53

This procedure is called a "wet." Now place the mirror blank with the beveled side facing down on top of the tool, place the palms of both hands on top of the back of the mirror blank maintaining pressure in the middle of the mirror, then position the mirror between a quarter and halfway to one side of the tool in the overhang position and begin sliding the mirror straight forward and back with long strokes for a few moments. Periodically alternate the overhang from one side to the other. You will hear the grinding sound as you do this because the abrasive is actually cutting into the glass and removing tiny amounts of glass with each wet. Stop, turn the mirror blank counterclockwise a small amount, step slightly around the barrel to the left, and proceed grinding the mirror in a straight line forward and back in the new direction. Continue to do this making your way around the barrel until you feel the grinding action of the particles becoming weak because they have been reduced to a powder and the movement of the mirror has become more resistant. Remove the mirror by carefully lifting it straight up and then pouring a little water on the tool, replace the mirror, and continue grinding for a few strokes to clean out the muddy remains. Do this a couple of times, then add more abrasive and spray a little more water for another wet, continuing the same process. The reason for all the turning is that the law of averages will cause an equal distribution of glass being removed from the front surface of the mirror blank, and as was mentioned previously more glass is removed in the center than at the edges, ultimately creating a concave curved surface. We must spend several hours, average between three and six, doing this before the curve reaches the Sagitta calculated beforehand. Don't be in a hurry to accomplish this making up your mind that this is going to take time over several days to complete. Altogether, grinding and polishing a mirror from start to finish demands around a hundred man-hours to finish the job. In order to generate a spherical curve we will adjust toward the end of rough grinding by using shorter strokes with much less lateral overhang. As with everything really important and worthwhile in life, lots of time and patience are needed to achieve success.

Another simple method to check the amount of curvature is to hold the mirror with a wet surface outdoors on a sunny day and reflect the sunlight onto the side of a wall or building. The distance between the mirror and wall when the reflected sunlight is reduced to the smallest spot will give an approximation of focal length.

Figures 54 and 55 below show the process of grinding back and forth and turning periodically to catch other parts of the mirror's surface.

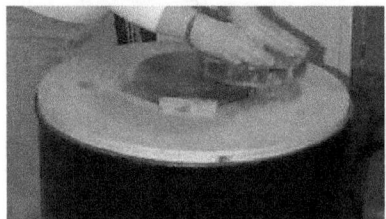

Figure 54

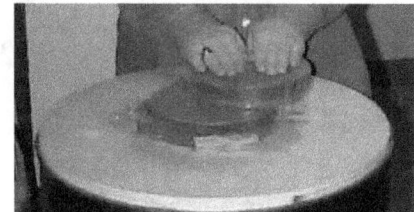

Figure 55

Figure 56 below shows a normal 'W' pattern that is a common technique used in the stroke by many mirror makers.

The 'Normal Stroke'

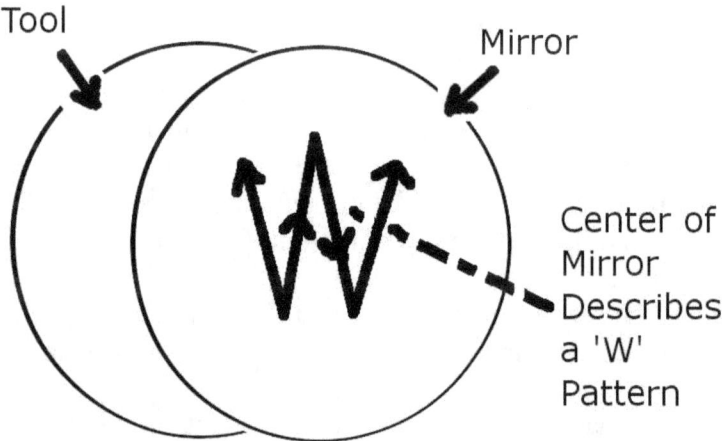

Figure 56

It's important to keep in mind that the only objective now is to cut away enough glass to set the curve we are interested in. Once accomplished, we no longer need to cut away large amounts of glass as this lofty goal has been achieved. In chapter 7, our focus will shift to smoothing the optical surface more and more until it becomes a thing of beauty.

Back to the Old Grind

Now that we have generated the mirror curve through rough grinding we must smooth the surface by a process known as fine grinding. Using the same technique we did in rough grinding, the difference with fine grinding is that it is done using a series of smaller and smaller abrasive grits. A large range of abrasive sizes will be used. The purpose of using successively finer grades of grit is to smooth out the pits and scratches left by the preceding, coarser abrasive grits. A total of approximately eight wets using each smaller grit size will work in removing the larger pits made by the previous one. When completed, wash the mirror under running water and stand edgewise where it won't fall to let air dry. Then hold the mirror up to an indoor light to examine its surface for any remaining large pits amongst the new finer pits. A magnifying lens can help here. We must be very careful not to allow the different grit sized abrasives to become contaminated with another. When changing to a finer grit abrasive, it's a good idea to thoroughly clean the work area and even wash your clothes so any contamination doesn't take place. Of course, both the mirror and tool should be washed completely. As we progress on to the finer grit abrasives, there will be a tendency for the mirror and tool to stick together. Keeping each wet fresh and using shorter fast strokes with the mirror nearly centered over the tool will keep this to a minimum (Fig. 57-58). If they do get stuck soak them in warm water for a while to loosen so they will come apart more easily.

It's important to avoid scratching the glass mirror's surface by spreading and distributing the fine abrasives across the tool's surface equally, not allowing the abrasive to form lumps. We can do this with our fingers and then squeeze out any excess water by pressing the mirror down on the tool before starting to grind. Use care around the edges of both glass disks and keep from drying. Also, periodically check that the focal length of the mirror has not been altered too much. If it becomes too short, then simply reverse by grinding with the tool on top over the mirror having the same curved surfaces in contact as before. As you approach the last stages of fine grinding, an adjustment to our stroke can make a difference in the curve of our mirror. By adding a little more pressure to the part of the mirror that is opposite the part overhanging the tool, the curvature will alter from spherical to that of a parabola. Ultimately, we want to parabolize the surface to avoid a condition known as spherical aberration, where the different frequencies (or colors) of light focus at varying lengths and the resultant image appears blurry. This was the problem associated with the Hubble Space Telescope upon its first launch into space. Also, lukewarm water should be used during the last wets of fine grinding to prevent any shrinkage of the mirror face and consequently grinding any of the edge away.

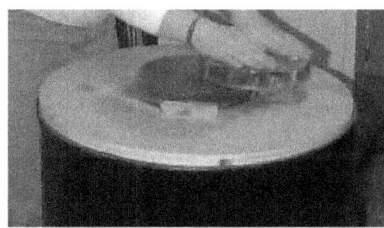

Figure 57

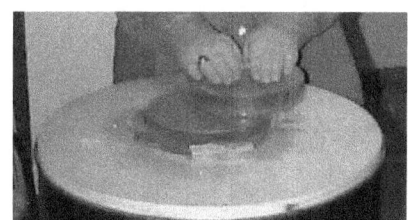

Figure 58

Figures 59 and 60 below show how the abrasive can tend to lump together, in which case we can use our finger to spread them around more equally.

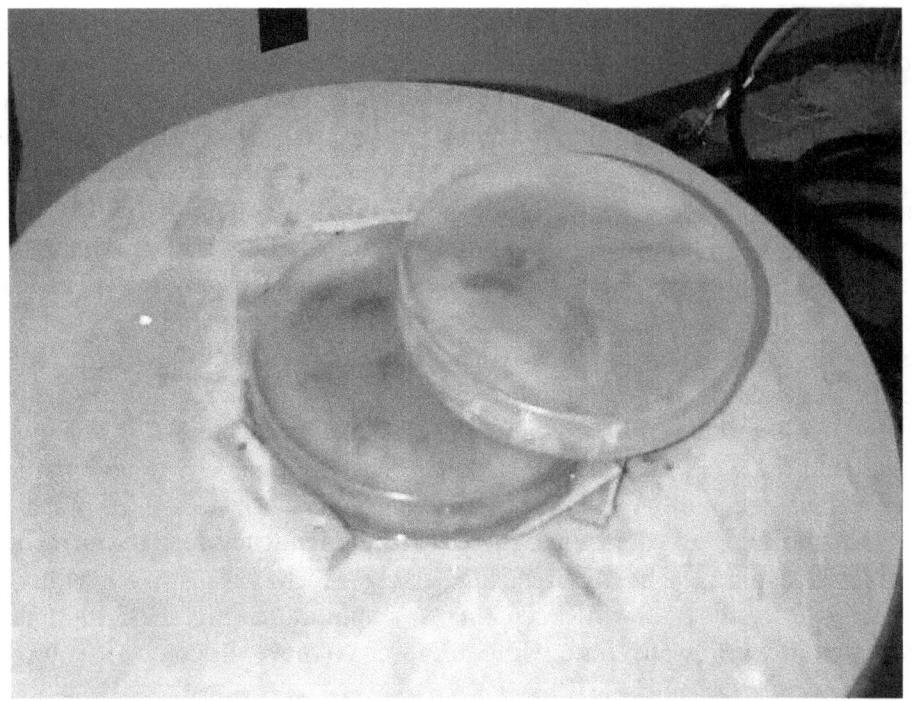

Figure 59

Figure 60

As we approach doing the final wets, adjusting to shorter and faster strokes with less overhang over the tools edge will help prevent sticking, as seen in Figure 61.

Figure 61

Applying more preesure on the mirror opposite the overhanging side will reshape a spherical curve into a parabola, as seen in Figure 62.

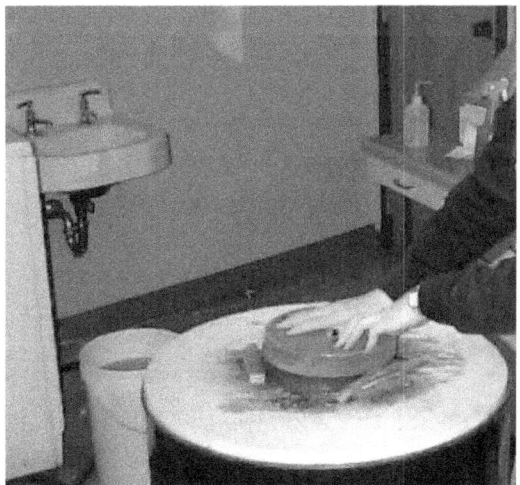

Figure 62

At the end of fine grinding, the surface of our mirror will shine with a beautiful glow of smoothness and crystal clarity. But we're not done quite yet. Now we move on to the final stage where the surface accuracy and smoothness of our mirror will far exceed what the eye can see, and an attempt toward optical perfection is the order of the day. In chapter 8, we will focus on polishing our mirror.

Groomed and Well-Polished

So far we have produced a curve in our mirror through rough grinding and smoothed it through fine grinding. Now we are ready to complete the smoothing process by polishing the surface of our mirror. Some errors during fine grinding can lead to problems as we attempt to polish our mirror. For example, if you didn't shorten the stroke length or the amount of lateral overhang during the fine grinding stage, you will find that the central region of the mirror is difficult or impossible to polish during the polishing stage. If you put too much pressure on the overhanging edge of the mirror, a nearly one inch diameter hazy ring may show up near the edge which will also not polish. Also, astigmatism is a defect that occurs when differing pressure is applied at right angles causing the curve to be slightly steeper on one axis compared to the other. There are other causes for astigmatism to occur as well. The art of the optician's craft in hand-figuring a parabolic mirror becomes refined with experience.

Creating a Pitch Lap

Pits form on the surface of the mirror while grinding from the abrasives rolling between the mirror and tool and chipping small pieces of the glass. The sizes of the pits are reduced with the use of successively smaller grains of abrasive during fine grinding. In polishing, a pitch lap is used to create a softer and less rigid surface in which very fine abrasive, such as white aluminum oxide lapping powders and fast polishing cerium oxide, are embedded to gently polish the surface. Tar or bee's wax can be used as pitch but we will be using the tempered burgundy pitch provided in our kit.

Place the pitch into a clean coffee can and slowly heat over a stove top and stir until it is well-mixed and yet thick and viscous when poured. See Figures 63 and 64 below.

Figure 63

Figure 64

As with each transition using finer abrasives during fine grinding, we must thoroughly wash and rinse the surfaces of both our mirror and tool using warm water only. Newspaper sheets can be placed over the work surface. Place the tool curved surface up, wet the surface, and spread some white aluminum oxide evenly over the entire surface. Do the same with the mirror surface. Carefully pour the hot pitch on the center of the tool face until it spreads evenly and flows close to the edge. Keep the pitch from reaching the edge where it would roll off. You can use your wet fingers to do this. See Figures 65-67 below.

Figure 65

Figure 66

Figure 67

Once the pitch has reached the point where flow has come to a stop, very gently press the wet mirror face curved surface down on the pitch until it comes close to the edge between the mirror and tool. To keep the mirror from sticking to the setting pitch, twist it and remove from the surface of the pitch quickly. This process has provided a polishing surface for the mirror. See Figures 68 and 69.

Figure 68

Figure 69

Immediately take the pitch lap tool and press into the pitch to form grooves perpendicular to each other slightly off center in the surface of the pitch (see Figure 70 below). This will produce a checkerboard or waffle-like appearance of squares across the surface. Just like grooves are cut into road surfaces to prevent cracking during freezing, the grooves in our pitch lap are to allow even pressure while conforming to the face of the mirror. See Figures 71-73.

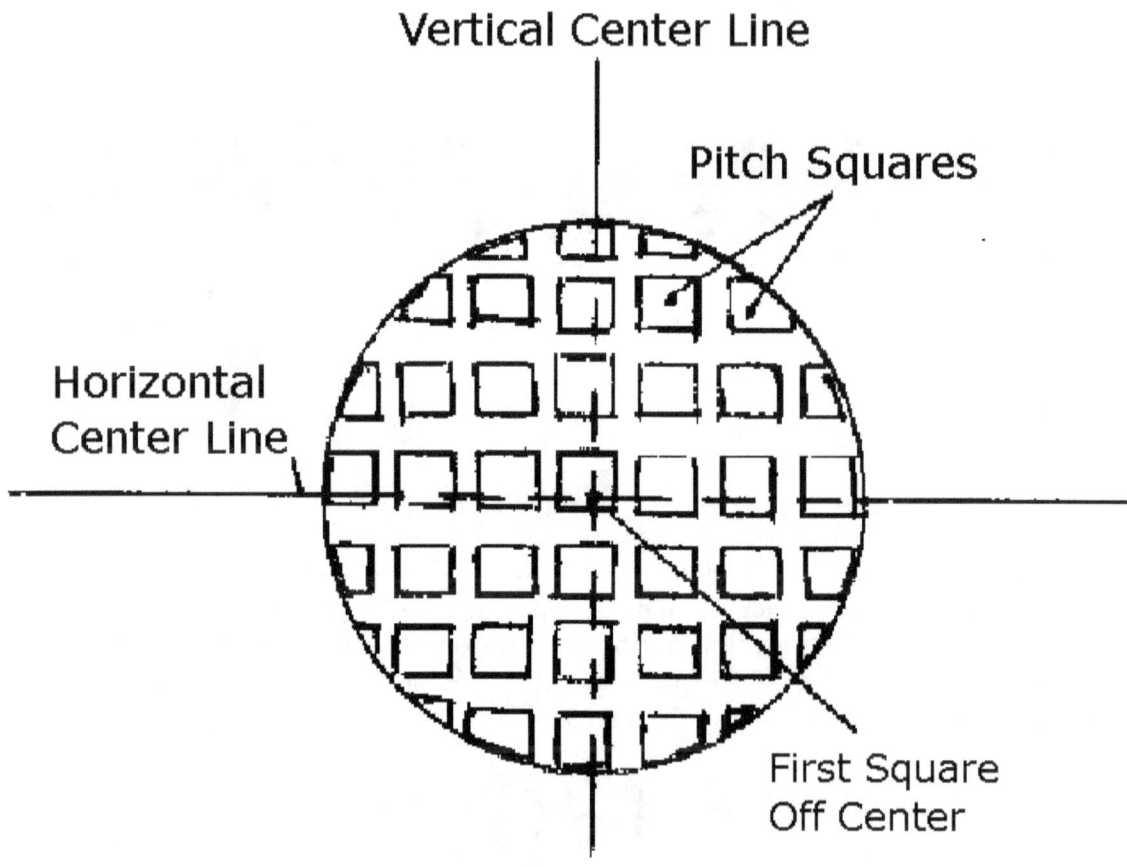

Figure 70

Figure 71

Figure 72

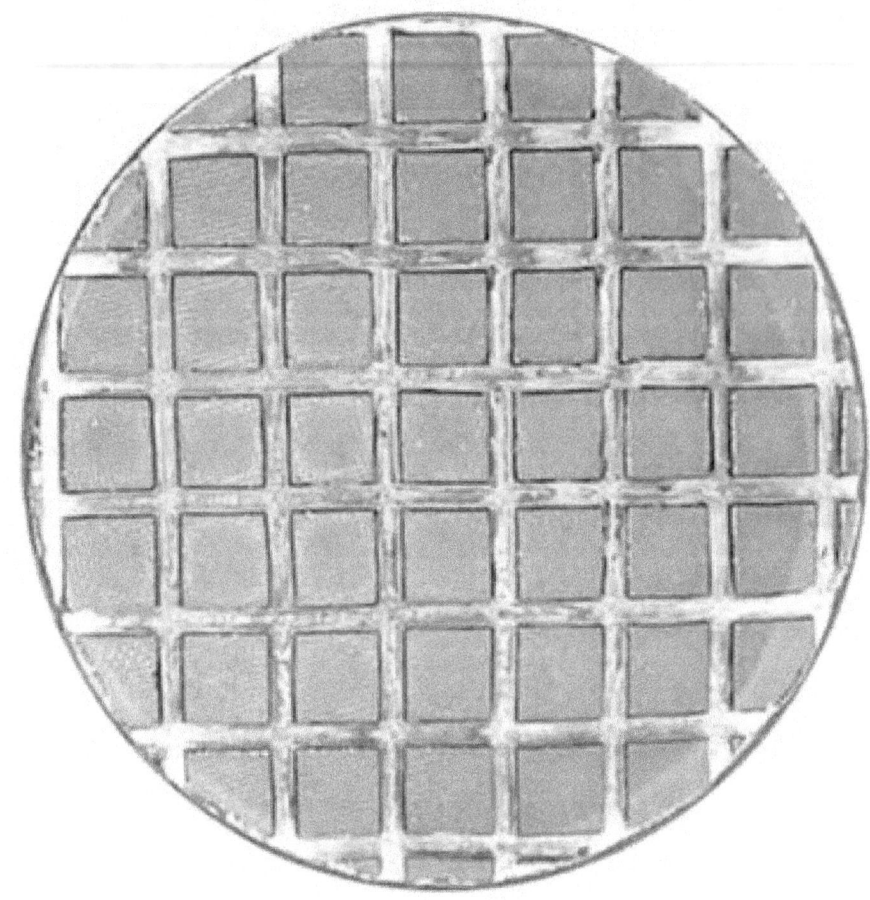

Figure 73

Once again press the mirror down onto the newly formed pitch, which should be hard enough now that the grooves are not compromised by pitch flowing into them. If this is not the case, refrain from doing this step until the pitch has set a bit more. You can repress the pitch lap tool if needed. Let the mirror sit on top while the pitch cools and sets, removing excess pitch that has overflowed the edge and gently moving and turning the mirror occasionally for equal heating. Once the pitch lap is cooled and firm, polishing can begin.

Polishing

Ideally, polishing should begin once the pitch lap is firm and continued until finished. Spray some warm water on the pitch lap and evenly distribute some cerium oxide onto its surface. While polishing, we will be using strokes similar to when we were fine grinding toward the end. It is important throughout this process to keep the pitch lap conformed to the mirror's surface at all times. Running cool water through the grooves and using a rubber mallet or your hands can help loosen the mirror if it begins sticking to the lap. If you still find the mirror is stuck to the pitch lap, you can soak it in warm water and use a pipe clamp fixture to separate them, as seen in Figures 74-76.

Figure 74

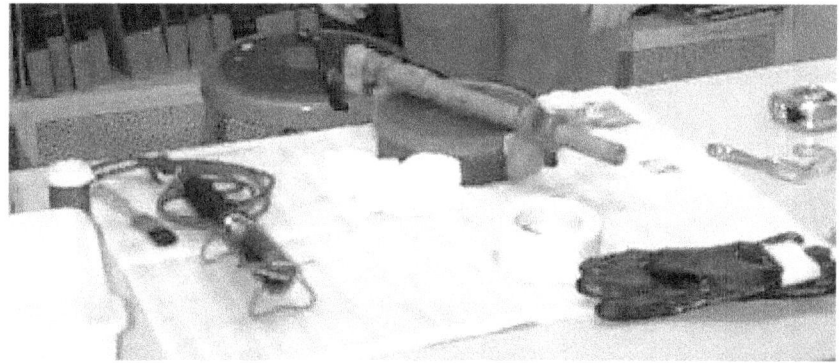

Figure 75

Figure 76

If you must leave part of the polishing for later, make sure to air dry both surfaces before placing the mirror down on the pitch lap to retain conformity. Keep the pitch lap clear of dirt or anything that would offset the perfect contact between the mirror's surface and pitch lap. Gently scrub with a clean brush under running water if this happens. Because of the close contact and small size of the cerium oxide, we will find that the mirror is more difficult to move while stroking. We can add some more cerium oxide to help prevent sticking. The total time needed for polishing will approximate 1 ½ - 2 hours.

Always begin polishing with short gentle strokes until the mirror has had a chance to warm up. The pitch lap with cerium oxide doesn't cut the glass as much as it does in smoothing the surface of the mirror. Eventually, the surface will change from hazy to crystal clear (see Figure 77 below). Then we are ready to work on "figuring" the mirror to a parabolic curve down to an accuracy of 0.000002 inch.

Figure 77

Figuring the Mirror

The process of "figuring" a mirror is to shape the optical surface to a paraboloid so it focuses all frequencies (colors) of light at the center of curvature. One simple method used to "read" the optical surface is commonly called a knife edge or Foucault test. This design was invented by the French scientist Leon Foucault in 1858. It uses a metal razor blade to create a point source of light placed near the center of curvature of a concave mirror and set up on a table top with the mirror standing vertically in a support at one end of the table. A long table that is a good choice for this project is shown in Figures 78 and 79 below. The Foucault test equipment used is shown in Figure 80. A diagram labeling the parts is illustrated in Figure 81.

Figure 78

Figure 79

Figure 80

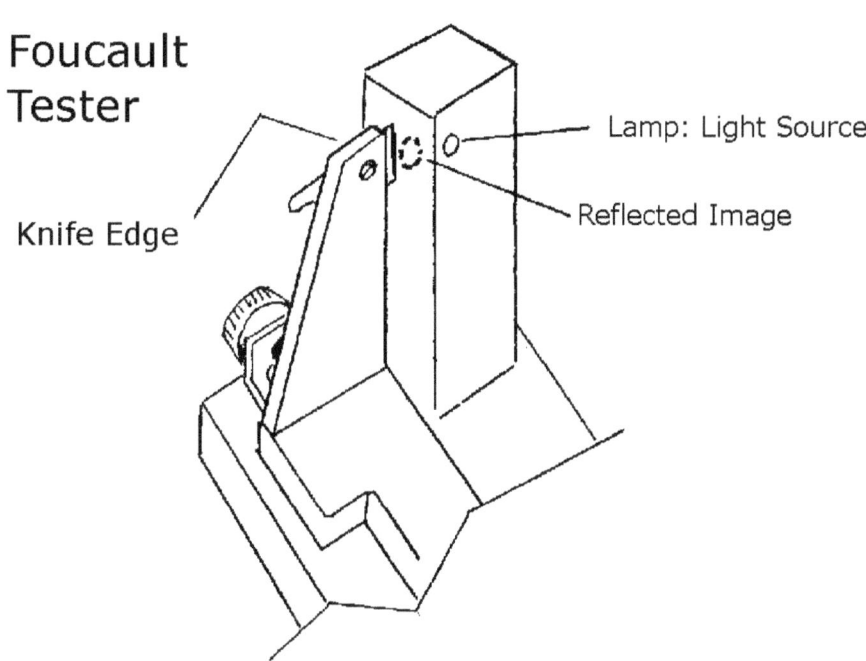

Figure 81

Essentially, the figuring process consists of testing the curve for defects and re-shaping it by altering our polishing strokes to correct the surface. Testing methods we will use include both the Foucault test (resulting images seen in Figure 82) and completing the optical tube assembly described in the next chapter so the test can be made at the focus, where the eyepiece of the telescope is. This latter method is commonly called a star test. By focusing the telescope on a fairly bright star and then adjusting the focus both inside and outside of this focus so the star's light is distributed in each out-of-focus disc, we can tell if the discs appear similar or if they are not identical, what the problems are that need to be corrected. If the defects in our curve reveal that it appears dark in the middle of the out-of-focus disk outside of focus and bright in the middle inside of focus, then we know that the center of curvature in our mirror is too deep (see Figure 83). This problem is remedied by using short strokes with the mirror centered over the tool. If the opposite is the case, appearing bright in the middle of the out-of-focus disk outside of focus and dark in the middle inside of focus, then the center of curvature in our mirror is too shallow. This is remedied by using long strokes with the mirror hanging over the edge of the tool, being careful not to damage the edge of the mirror. If the outside focus of the star shows a bright ring at the edge of the disc, this indicates that the curve is too flat near the edge, commonly referred to as a "turned down edge." This defect can be remedied by overhanging the mirror halfway over the pitch lap and applying more pressure at the edge of the mirror over the center of the tool.

Foucault Test

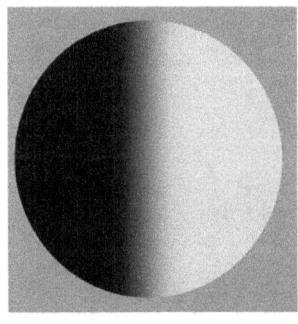
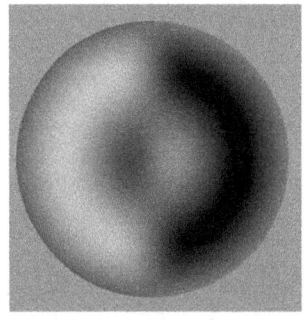
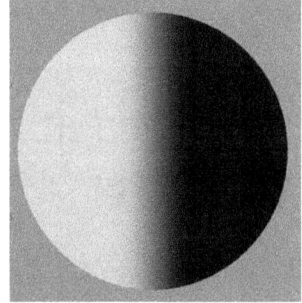

| Inside focus | At focus | Outside focus |

Figure 82

Star Test Showing Center of Curve Too Deep

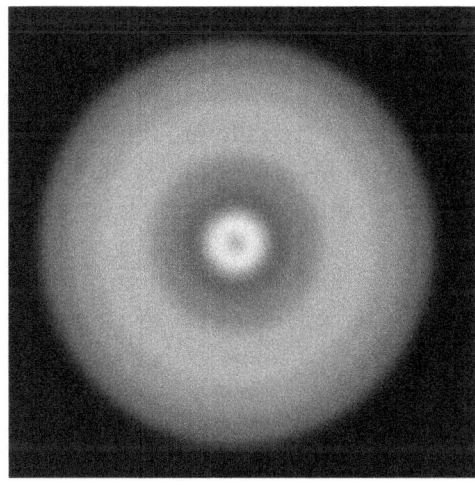

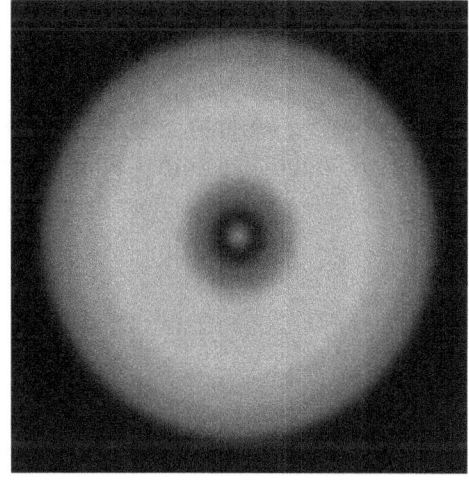

Inside focus   Outside focus

Figure 83

Once you are satisfied with the result of correcting defects on the surface of your mirror, you will then want to aluminize the surface of your mirror with a vacuum chamber to provide a durable reflective surface. Some optical companies listed in the next section can provide this service for you if you send the mirror or bring it to them.

Finished off-the-shelf parabolic primary mirrors are available for purchase should you choose to skip this step and concentrate your efforts on building the rest of the telescope. One of the most reputable optical companies, Parks Optical has been offering superior quality telescopes and accessories since 1954. Their primary mirrors are downright expensive, but well worth the price if affordable and within your budget. You can see what they offer at http://www.parksoptical.com. Another company offering outstanding quality optics is Discovery Telescopes, Inc. at http://www.discoverytelescope.com/optics.html. You can order by phone by calling 1-877-523-4400 or 949-347-0142 and ask for Bill Larsen. If you'd rather keep costs down without giving up great quality, then Orion Telescopes & Binoculars at http://www.telescope.com may be the direction to go in. If you don't want the expense that new primary mirrors can demand, used telescope mirrors can be purchased through the classified ads listed in the online AstroMart at https://astromart.com/classifieds/astromart-classifieds/telescope-making/show or TelescopesForSale.org at their web site http://www.telescopesforsale.org/telescope- accessories/telescope-mirrors/, and of course there are always lots of telescope parts for sale regularly on EBay.

The Tube Assembly

The optical tube assembly basically holds and aligns the mirrors of our reflecting telescope so they can bring light to a focus for viewing at the eyepiece. At one end of the tube we have our primary mirror mounted in a mirror cell, at the other end the diagonal mirror mounted in a diagonal mirror holder held in position by a 4-vane spider and an eyepiece focuser. The tube itself is painted inside and out and has end rings for support and trunnions mounted at a balance point fitted to the rocker box so the tube can point up and down. Finally, a finder telescope is mounted near the top where the eyepiece focuser is located. Now let's see how each of these parts comes together.

I used to recommend Sonotube concrete forms made by Sonoco construction products and are used to set concrete pillars. They used to make excellent tubes for telescopes because they were rigid, lightweight, paintable (wax film inside must be removed), and come in 2-inch increment diameters from small to large. Now, they are flimsy thin-walled tubes that are not good for a telescope. Now I highly recommend Yazoo Mills, Inc. because they can make any wall thickness you want. Go to https://www.yazoomills.com/ to see their website, or call them at (717) 624-8993. It must be cut to the proper length to support the optics at their focus but also have a few extra inches at top for keeping out stray light and dew. The size we need for this telescope is a 10-inch tube. It has an inside diameter of 10 inches and an outside diameter of 10 ¼ inches. You will need to cut the tube to the proper length for your telescope's focal length. For my 8-inch f/8 reflector with a focal length of 64 inches, I cut the 10-foot long tube I bought to 66 ½ in. long (67 ¼ in. with mirror cell). See Figure 84 below for setting the distance from the rip fence to the blade (side facing rip fence) using a tape measure.

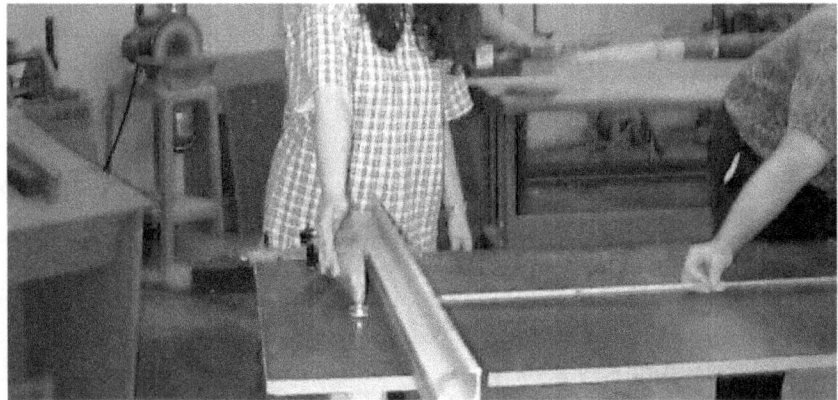

Figure 84

Attach a straight wooden brace (such as a 2x4) along the length of the table (and tube) near the blade on the side it is cutting toward. With the blade guard removed, set one end of the sonotube up against the rip fence brace so that the tube is on the other side of the blade in a straight line with the table. Being very careful, turn on the saw and slowly roll the tube toward you so the blade begins cutting the bottom of the tube. Once the tube hits the wooden brace, slowly rotate the tube so the top rolls toward you and the bottom away into the blade for a continuous cut all the way around. It's important to keep the end of the tube pressed into the rip fence at all times for an even cut. I found it very helpful and practically mandatory to have other hands helping with this process. For safety, all hands must be kept away from the saw blade at all times. See Figures 85 and 86 below.

Figure 85

Figure 86

Once the tube is cut to the proper length, the waxy film coating must be removed from the inside of the tube. Many years ago I contacted the owners from some reputable companies to inquire how to remove this wax film. One told me to sand it out using sand paper. It took a very short time to realize that this procedure could take a lifetime to complete. Another "expert" advised me to torch it out. But I thought torch, spiral paper, not a good combination. I found for myself that if you pinch the tip end of where the wax film spirals to a point at the end of the tube, pull up slowly, and continue to unravel the wax film from off the inside wall of the tube as far as you can reach on one end, you can then begin the same way from the other end and reach the middle where all of it can be pulled out (see Figure 87 below).

Figure 87

With the wax film removed, paint will adhere to the inside surface of the tube. Since these tubes are typically used in construction to set concrete pillars, the waxy coating inside helps remove the spiral paper once the concrete sets. Using the small 3-inch heavy matte roller with handle sold at any paint store or department, generously coat the inside of the tube with flat black paint. Flat black is used as an anti-reflective coating to prevent stray light from being scattered inside the telescope's light path. Since the handle will have threads to screw an extension handle (broom handles work fine), we can easily reach the middle of the tube with the flat black paint from both sides. See Figures 88-94 below.

Figure 88   Figure 89

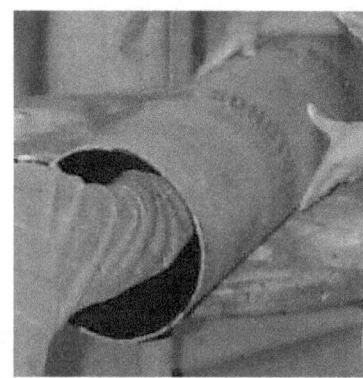

Figure 90   Figure 91

Figure 92

Figure 93

Figure 94

Now that we have finished painting the inside of the tube, we can choose a color to paint the outside of the tube with. Since telescopes are used at night when it is dark, a light color is preferred. I chose white since it would show up best in the dark. Semi-gloss house paint with a brush works fine. See Figures 95 and 96.

Figure 95

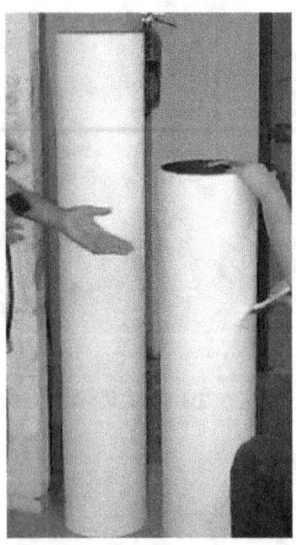

Figure 96

Tube end rings serve as reinforcements to both ends of the tube, a place to attach other equipment, and something to grasp better when repositioning or tracking celestial objects. Using the square cutouts from our original sheet of plywood, we are going to cut perfect circles a few times. A router works best for this, but I chose to use a jig or sabre saw to make the cuts. On one side of the saw, I attached a piece of hardwood perpendicular to the blade's cut using a wood screw. Then I drilled a hole through the hardwood so the <u>inside</u> of the blade would exactly equal the radius of the circle's outside diameter of 13 ⅛ inch. I also drilled a hole in the center of the square plywood cutout about ⅝ inch down so a steel pin could hold the hardwood in position, serving as a guide for the saw to pivot around cutting a perfect circle (counterclockwise when viewed from above). Placing the two opposite sides of the board on top of separate tables, I now had the clearance to make the cut without obstruction (see Figure 97 below).

Figure 97

Then I drilled a hole through the hardwood so that the <u>outside</u> of the blade would exactly equal the radius of the circle's inside diameter of 10 ¼ inch (the outside diameter of the sonotube). Again, placing the edges of the circle on top of the edges of two tables the right distance apart enabled me to make the cut freely. See Figures 98 and 99 below.

Figure 98

Figure 99

The aftermath of all the original circular cuts can be seen in Figure 100.

Figure 100

Now we have solid oak plywood end rings to attach to each end of the tube. Figure 101 below shows the result.

Figure 101

Sanding both the inside and outside cuts of our rings will prepare them for a coat of varnish as seen in Figures 102 and 103.

Figure 102

Try fitting them over the ends of the tube first, however, and sand as needed to make them press fit snugly onto the ends of the tube (Figures 104 to 106). You may need a wood mallet to gently hammer them in position.

Figure 103

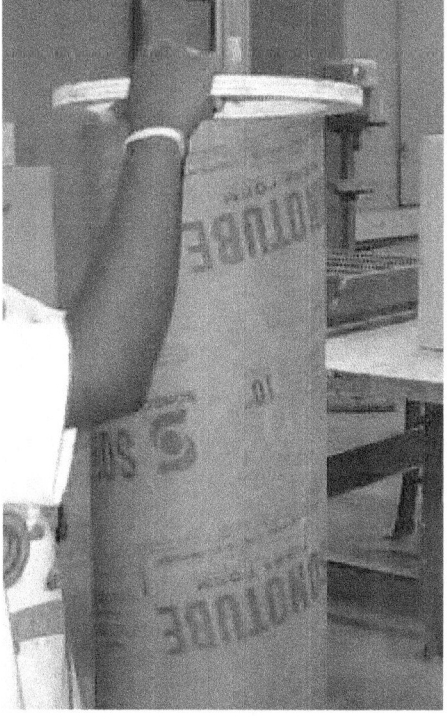

Figure 104

Figure 105

Figure 106

Coat each ring with sanding sealer, let dry and lightly sand, then brush a final coat of varnish for a glossy and protective finish (Figures 107 and 108). Now drill four small holes equidistant from each other in the center of one ring slightly larger than the outside diameter of the threaded part of tee nuts for screwing machine screws into in order to attach the mirror cell to this "back" ring (Figure 109). Hammer the four tee nuts securely into each of these holes. With the tee nuts facing upwards toward the tube, attach this ring at one end of the tube using twelve 1 ¼ inch long hex head wood screws. Attach the other "front" ring to the opposite end of the tube the same way (see Figures 110 and 111).

Figure 107

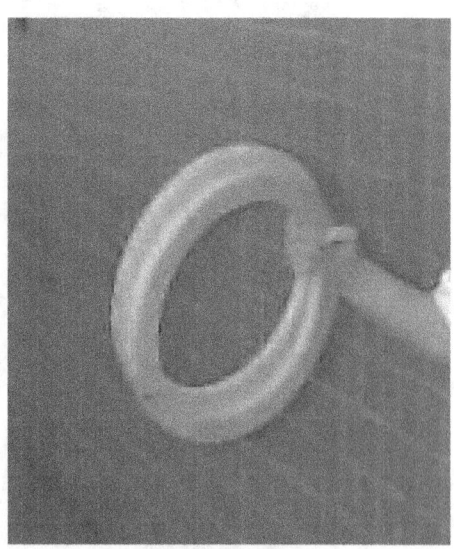

Figure 108

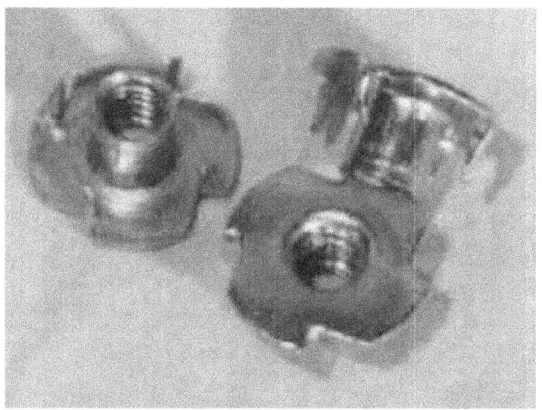

Figure 109

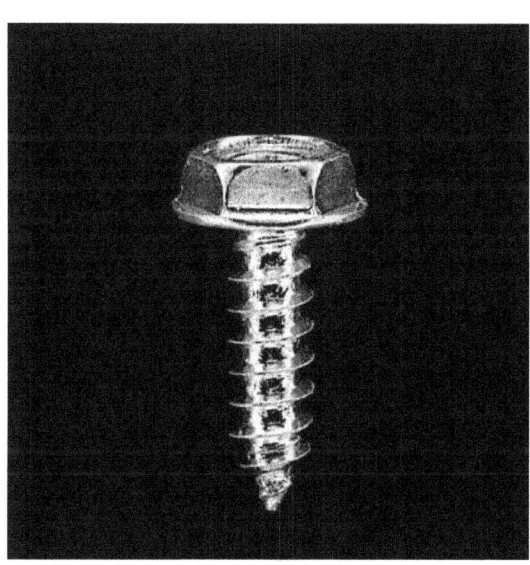

Figure 110

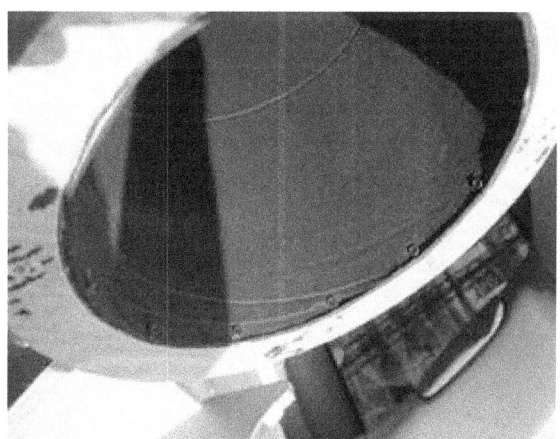

Figure 111

The third circle is cut the same size as the outside diameter of the end rings (13 ⅛ in.) and will form part of the mirror cell that holds the primary mirror in place at the back of the tube. Four 2-inch diameter holes need to be drilled through this circle so air can flow freely through the telescope tube and allow the mirror to adjust to ambient, or surrounding outside temperature. The reason for this is that until the mirror cools down, convection currents can form distorting the image like a mirage over a highway, and the sooner it cools down the less time we need to wait for observing. Using a tape measure, protractor and a pencil, mark one dot in the center of the circle and three dots each 4 inches from the center and 120 degrees apart. Each dot marks the center of where a 2-inch hole will be drilled. The outer perimeter of the holes will be inside the tube itself when attached to the back end ring. See Figure 112.

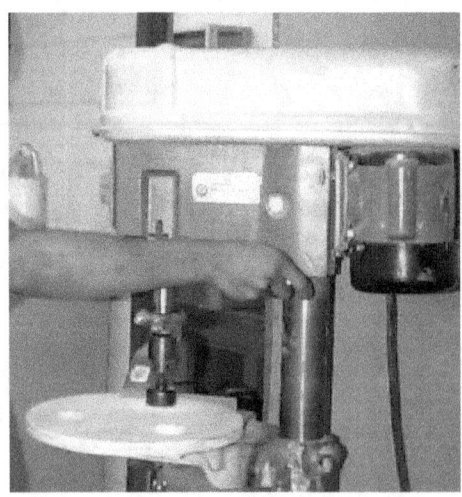

Figure 112

After the holes have been drilled (Figure 113), sand all surfaces, aply sanding sealer, sand again lightly, and paint a final coat of varnish, making sure to paint one side (facing up into the tube) and the inside rims of the holes flat black to minimize reflection (Figure 114).

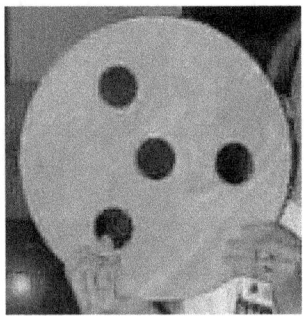

Figure 113

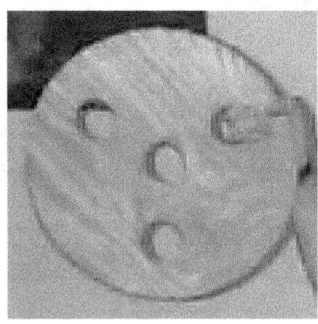

Figure 114

An aluminum mirror cell can be made or purchased for holding the primary mirror (see Figures 115-117 below). Three small holes slightly larger than the collimating bolts of the aluminum mirror cell now need to be drilled between the three 2-inch holes in our circle (Figure 118).

115

116

` 117   118

Three compression springs surround each bolt, and one end is in contact with the plywood circle facing toward the tube while the other end is in contact with the back of the mirror cell dish (Figure 119). This way, when wing nuts are tightened or loosened on the plywood circle facing away from the tube, the mirror is tilted at three points allowing its reflective front surface to focus down the precise center of the tube (Figure 120).

119   120

Finally, four small holes must be drilled equidistant around the outer perimeter of the plywood circle for machine screws to screw into the tee nuts on the tube-facing side of the back end ring to hold the mirror cell in place (see Figures 121 & 122 below).

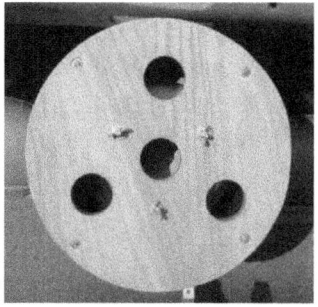

Figure 121   Figure 122

In order for the optical tube assembly to be able to swivel up and down in altitude pointing to celestial objects higher or lower in the sky, we need to attach parts that will serve as trunnions allowing it to pivot on top of the rocker box. Let's start with a simple rectangular 12 in. x 6 in. cut of plywood to act as a bracket connecting the other pieces as a unit. Since we need one on each and opposite sides of the tube, two of the same will need to be used. See Figure 123 below.

Figure 123

For the most secure way to mount these to the sides of our tube, we can attach end contacts with the curve traced from the outside perimeter of the tube itself. With one on each end of the bracket, we'll need a total of four identical pieces. Figure 124 shows one of these end contacts. For proper spacing within the fork arms of the rocker box, a thickness of ⅞ inch in the middle of the curve will do.

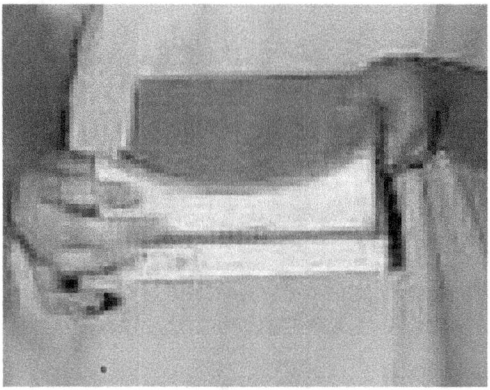

Figure 124

The last part of our trunnions are the most important as they will be acting as a bearing surface for the pivot action of the tube. Two 7-inch diameter circles will work. We don't want the brackets rubbing inside the fork arms of the rocker box, so adding two additional 7-inch circles cut from ¼ inch plywood glued to the ¾ inch circles will make a full 1-inch thick circle with ¼ inch clearance on each side and even with the outside edges of the rocker box. Figures 125 and 126 show each of these circles.

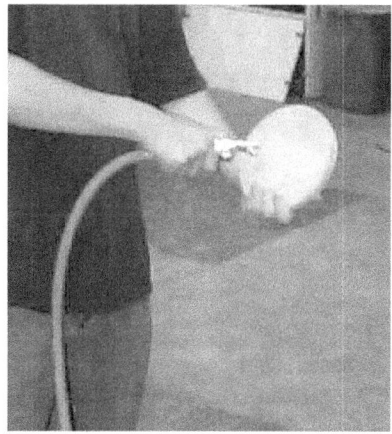

Figure 125

Figure 126

To reduce unnecessary weight, we'll drill 2-inch holes in the center of each of the circles. Figure 127 shows the finished circles for one side after they've been sanded.

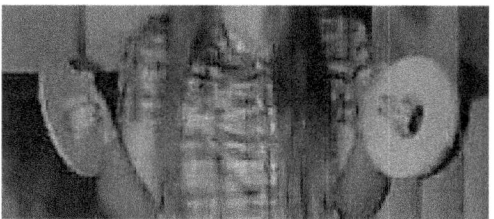

Figure 127

Using our Tightbond wood glue and drywall screws, let's assemble the parts as a single unit, as shown in Figure 128.

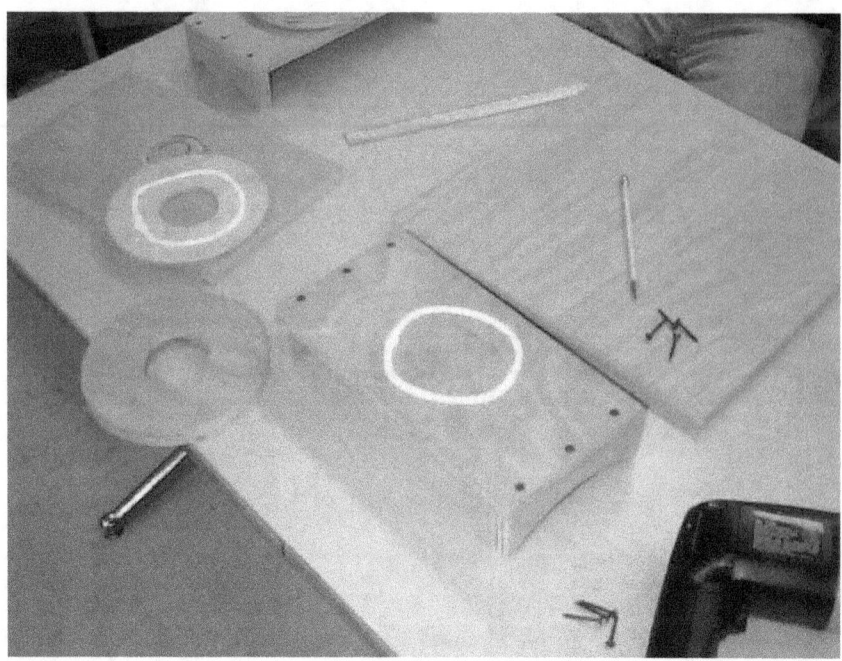

Figure 128

The finished product is pictured in Figure 129, which after applying sanding sealer, sanding, and adding an outer coat of varnish to all surfaces of both units except the round edge of the circles (Figure 130), we are ready to use contact cement and add our strips of Ebony Star Formica around the edges of both circles. Finally, we can mount them to opposite sides of the tube using three 1-inch long hex head wood screws through the inside of the tube into the four curved parts of the end contacts. One easy way of ensuring proper balance and perfect opposite side alignment, wait to do this until all other parts, including the mirrors, focuser and finder scope, are all in place in the tube. These parts will be mentioned later in this chapter. Then through trial and error, hold the brackets on each side of the tube so the focuser is at a 45-degree angle between straight up and facing right from the rear, and allow friction to work as you carefully place this over the Teflon pads of the rocker box. Adjust until a good balance is achieved with the tube in different high and low positions, trace a line at the inside and outside edges of each end contact onto the tube using a pencil, and drill six holes slightly larger than the shaft of the screws not counting the threads through the tube wall, being careful to avoid the center and edges. **Make sure to take the mirrors and other supportive equipment out before doing any drilling**, not only to allow room to do so, but most importantly to protect the optical surfaces from particles striking them. Hold one of the completed trunnions at a time in place within the lines drawn, and drill holes into the wood from inside the tube. Now screw into each of these predrilled holes from the inside of the tube using the 1-inch long hex head wood screws. The tube and trunnions will now become a single solid unit.

Figure 129

Figure 130

Ebony Star Formica strips like that shown in Figure 131 and 132 cut to the right diameter can be contact cemented around the circles' edges of our trunnions. Before applying the cement, wrap the strip around, mark the overlapping edge, and cut so no overlapping material will be there when permanently cementing these to the wood. A scissors or carpet knife can be used for the cut.

Figure 131

Figure 132

Figures 133 & 134 below shows the trunnions mounted on the tube with the Ebony Star Formica strips contact-cemented around the circles and sitting on two Teflon pads on each side of the rocker box. The amount of friction is just right for the telescope tube to pivot up and down almost effortlessly and stay in position once moved.

Figure 133

Figure 134

Brass screws can be used to secure the ends of the strips to ensure they stay in place, as seen in Figures 135 & 136.

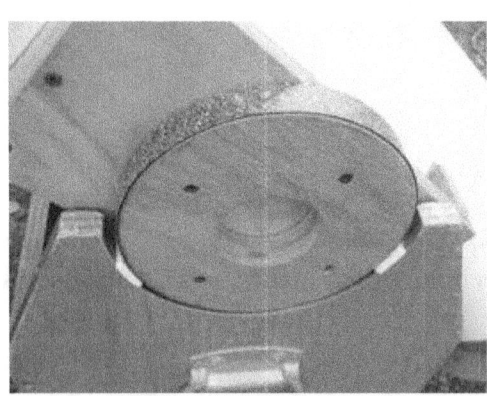

Figure 135

Figure 136

Another part of a complete optical tube assembly is a 4-vane spider and elliptical diagonal mirror holder. The latter holds the diagonal mirror at a 45-degree angle to the light path in the tube, while the spider simply holds this mirror in the exact center of the tube. These two devices work together as a unit and is pictured in Figure 137 mounted at the front end of the tube. The three screws are used to collimate the elliptical diagonal mirror to send the light path from the primary mirror directly into the focuser drawtube. Look also at the other side where the face of the diagonal mirror can be seen directly (Figure 138) and in the reflection of the primary mirror (Figure 139).

Figure 137

Figure 138

Figure 139

The four thin steel vanes of the spider held edgewise to our light path cause some diffraction (light going around the edge of a barrier) but minimal obstruction. It's the reason we see four spikes on many bright star images photographed through telescopes. The diagonal mirror and holder form a central obstruction to the light path, which can also be minimized by choosing the smallest one possible. Since this mirror has an elliptical shape, the smaller or minor axis is what will determine the diameter of our central obstruction. For our 8-inch mirror, a diagonal of 1.52 inch minor axis is typical and acceptable.

Before drilling the four holes near the top of our tube for the screws that attach the spider to the inside of the tube, we must calculate it's position based on the focal length of the primary mirror. To do this, start by measuring the distance of the front reflective surface of the primary mirror from the back edge of the tube. This will mark our starting point. If our mirror is an 8-inch f/8 like mine, then we know its focal length is 64 inches (8x8). Say the front surface of the mirror is 5 inches from the rear edge of the tube. The radius of the 10 ¼ inch outside diameter tube is 5 ⅛ inches, which marks the distance from the center of the diagonal mirror's surface to the outside edge of the tube. Let's also say our focuser stands 2 ⅞ inches above the surface of the tube. Now the light has to reach a distance of 8 inches from the diagonal mirror. If we subtract this 8 inches from the 64-inch focal length, we get 56 inches. That's the distance the converging cone of light from the primary mirror has to travel up the tube itself. Now we simply add this 56 inches to the 5 inches where the reflection began, then we know the hole for our focuser will be centered a distance of 61 inches from the bottom of the tube. Since the tube's length is 66 ½ inches, we can just subtract the 61 inch light path and know this hole will need to be drilled exactly 5 ½ inches from the top end of the tube. Now we measure the distance from the center of the diagonal mirror to the midpoint of one of its spider vanes (the position where the screw will secure it to the tube). Let's say that distance is 2 inches. So now we know that the screw holes for the spider are going to be 5 ½ – 2, or 3 ½ inches from the top edge of the tube. I like to place these screws at the top and bottom of the tube (as oriented with the trunnions) and then the sides. You can position these four holes accurately by using a flexible or sewing tape measure wrapped around the outside of the tube's perimeter and dividing by 4.

The rack & pinion eyepiece focuser (Figure 140) will be mounted directly over where the center of the diagonal mirror is below, at the point we determined previously as 5 ½ inches from the top edge of the tube. It makes good sense to position the focuser at a 45-degree angle for convenient viewing at all elevations. So if we measure halfway between the top and left side (as facing looking down the tube from its top), make a pencil mark and center it under the round drawtube of the focuser. Hold it steady while using a sharpened pencil to mark dots where the four screws will go to mount the focuser to the telescope tube. Drill holes where the four marks are and the focuser is ready to attach to the tube. Figure 141 shows the focuser completely racked out.

Figure 140   Figure 141

Finally, we must add a finder telescope, or simply finder scope (see Figures 142 & 143), to the outside of the tube for helping us locate and find celestial objects easily. Figure 144 shows the front view of the finder's objective lens and the instrument mounted at the top of the telescope tube. They help us find objects quickly by providing a large field of view of the sky and if properly aligned with the light path of the main telescope, we can center the object in crosshairs similar to the view in Figure 145. If you care to spend the money, an illuminated reticle device uses a battery to light and adjust brightness of four red lines (Figure 146). An inexpensive 6x30 finder is sufficient.

Figure 142

Figure 143

Figure 144

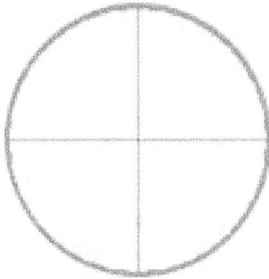
Figure 145

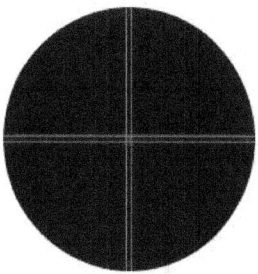
Figure 146

In Figure 147, we see the top inside end of the tube with spider and diagonal mirror and the bolts holding the focuser and the 6x30 finder scope in place. Figure 148 shows the finder and focuser attached from an outside view. Figure 149 shows the entire optical tube assembly completed alongside the finished rocker box and ground board, which are usually kept bolted together.

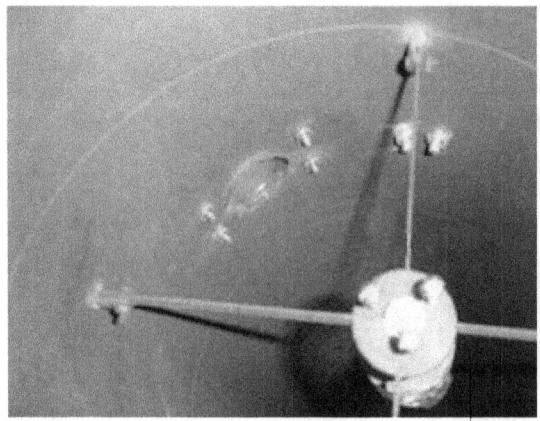

Figure 147

Figure 148

Figure 149

Finishing Touches

Okay, the glue has set and screws tightened, and all parts have been assembled together. The mirrors' surfaces shine with crystal clarity, and the only thing separating us from first light is a clear dark sky and a few minor adjustments. If we were successful in following each step with some degree of accuracy, then we can expect an excellent performing instrument that will give us pleasure for a lifetime.

The first chapter of this book, Telescopes 101, gave us the formula for calculating magnification for any given eyepiece and telescope combination. The theoretical limits of magnification is 3.5D and 60D, meaning that the lowest useful magnification is 3 ½ times the diameter of the telescope's primary mirror while the highest useful magnification is 60 times its diameter. So, for our telescope, the lowest power is 3.5 x 8 = 28X, and the highest is 60 x 8 = 480X. An eyepiece giving lower magnifications actually vignettes or effectively reduces the aperture of our telescope, resulting in less light gathering power. Higher magnifications can result in a fuzzy dim image with a loss of detail. It is typically referred to as "empty magnification" for this reason. What is found in practical experience is that much lower magnifications are usually preferred because of atmospheric turbulence and scattered light. The metric focal length of an 8-inch f/8 mirror is 1,625 mm. So a 40 mm fl eyepiece will give about 41X, and if it's a wide field eyepiece that would be a great low power one to have. A 16 mm eyepiece provides 102X, a good step up from the 41X and a good medium power choice. Add a 3X Barlow lens to the mix, and now our 40 mm gives us 123X and the 16 mm will yield 306X, a good high power for most purposes.

Before we can begin our first observation of the heavens, the optical system must be collimated accurately. We begin by looking down the drawtube of the eyepiece focuser and making sure the diagonal mirror is facing directly up through this opening. The shaft of the diagonal mirror holder is threaded so tightening the nut once turned up in proper position will hold it in place. Turning the entire diagonal mirror holder will adjust the distance toward or away from the primary mirror using the threads of this shaft, so position the diagonal mirror so it appears centered below the drawtube before tightening the nut. On the back or top-facing side where the nut was tightened, three screws will adjust the tilt of the diagonal mirror so the reflected light will come through and centered in the drawtube. Finally, adjust the three wing nuts at the back of the primary mirror cell until the reflected image of the diagonal mirror appears exactly centered in the round face of the primary mirror, as seen in Figure 150. The converging cone of light reflected from the primary mirror is essentially travelling up centered in the telescope tube, as viewed in Figure 151 without the diagonal mirror, holder or spider in place. There are tools available that help more precisely align the optics, but if you get the view down through the center of the focusing drawtube similar to that in Figure 150, you will find that to be sufficient for this longer-focus telescope. Shorter focal ratios, or faster systems, typically require stricter tolerances in collimation and are less forgiving. You will also notice a reflection of your eyeball in the middle of the diagonal mirror when doing this kind of alignment. Each and every time you set the telescope up for observing, always check the collimation of the optics and make any minor adjustments accordingly.

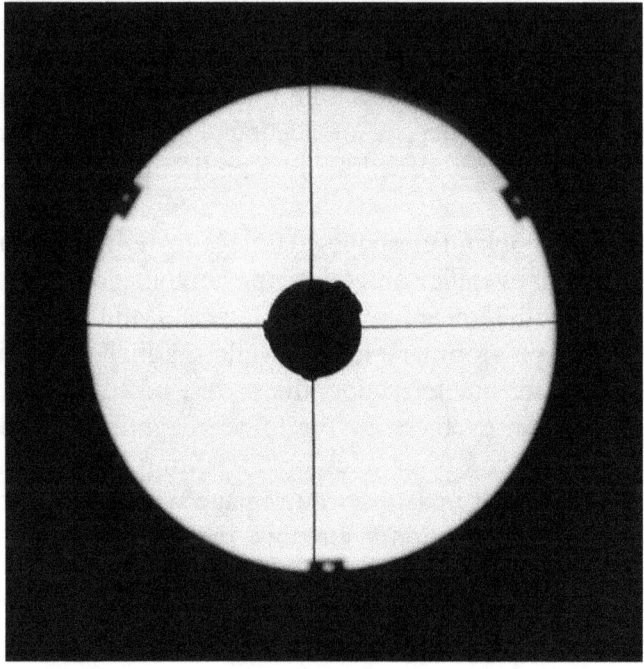

Figure 150

Figure 151

Some people prefer to use a curved spider vs. a 4-vane one upon claims that the image will be improved. It is true that the four spikes typically seen in star images disappears, but the diffraction is still there only it is more spread evenly throughout the image instead. For a seriously discriminating planet enthusiast, it may be the way to go. But for all practical purposes, a 4-vane spider will do fine. A telescope with one-half the normal curved spider is pictured in Figure 152.

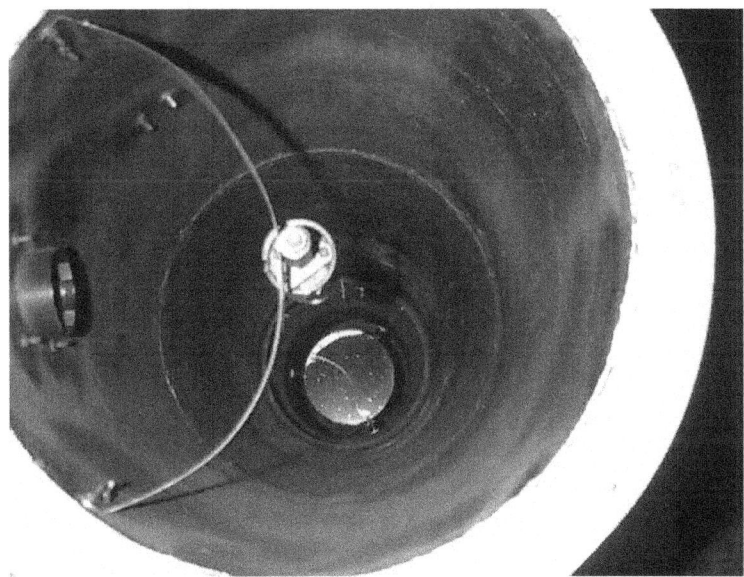

Figure 152

Another alignment we must do before using the telescope is that of the finder scope. This can be done most easily during the day. Using the low power eyepiece in the main telescope, center a terrestrial object such as a street sign down the block or the top of a tower, etc. Then while looking into the finder scope's eyepiece, adjust the three screws at the front end of the tube until the same object is centered within the crosshairs. Now double check that it is still centered in the telescope and you will be ready to easily locate objects in the night sky. A photo of our finished telescope ready for action is shown in Figure 153. Notice the brass collar at the bottom of the tube enhancing the telescope's appearance and adding a bit more weight at the mirror end so the length of tube below the trunnions is slightly shortened.

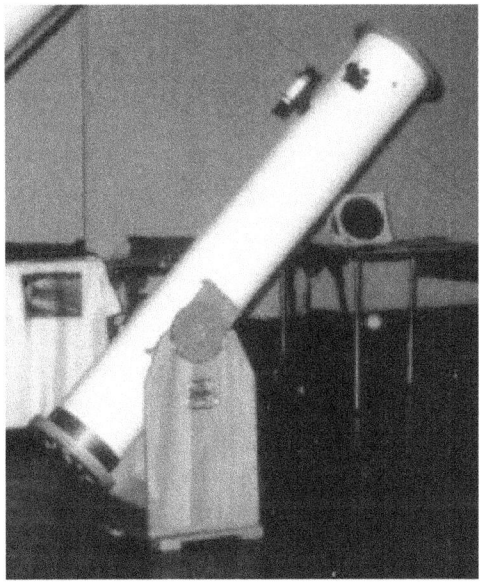

Figure 153

Down & Dirty

Our telescope is complete and now we are ready to put it into operation. Place the ground board on a relatively flat level surface and put the rocker box on top, positioning it so that the central hole aligns directly over the same for the ground board. Screw the center bolt into the tee nut below and hand tighten, connecting the two parts together as a unit. From now on we can keep these parts bolted together for transportation, storing, etc. See Figures 154-157 below.

Figure 154

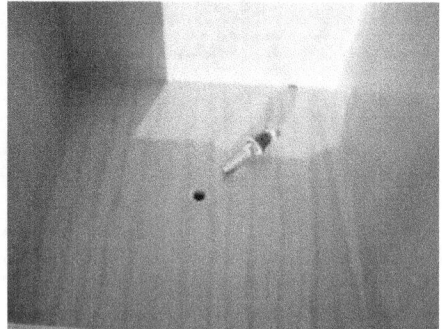

Figure 155

Figure 156

Figure 157

Place the optical tube assembly over the rocker box and carefully lower the trunnions on top of the Teflon pads within the semicircular cutouts of both sides. This is done, of course, with the focuser and finder scope facing up and the back mirror end of the tube faced toward the open side of the rocker box, as seen in Figure 158. A full view of the assembled telescope appears in Figure 159.

Figure 158

Figure 159

The next step is to look down into the focuser drawtube without an eyepiece to collimate the optics of our telescope. The view of the diagonal mirror should be centered below within the circle of the drawtube. If not, loosen the nut securing the bolt of the diagonal mirror holder and turn until the mirror becomes centered. Now check that the reflection in this diagonal mirror has a view of the primary mirror in the center. If not, loosen and tighten each of the 3 screws at the back of the diagonal mirror holder to adjust this tilt so that the image of the primary mirror is centered, as seen in Figure 160 below.

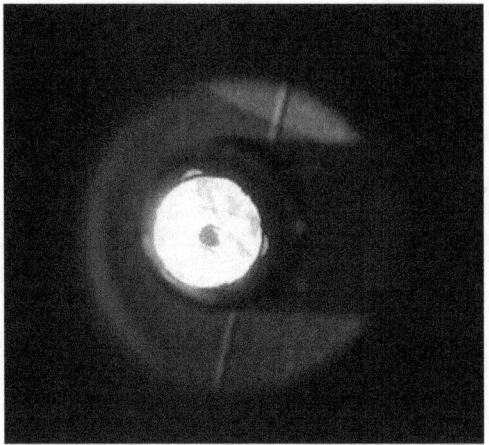

Figure 160

Finally, the reflection of the diagonal mirror in the face of the primary mirror should be centered. If it's not, then loosen or tighten each of the 3 wing nuts at the very back of the tube to adjust the tilt of the mirror cell. We should be able to see the reflection of our eye centered in the diagonal mirror. See Figures 161-165.

Figure 161

Figure 162   Figure 163

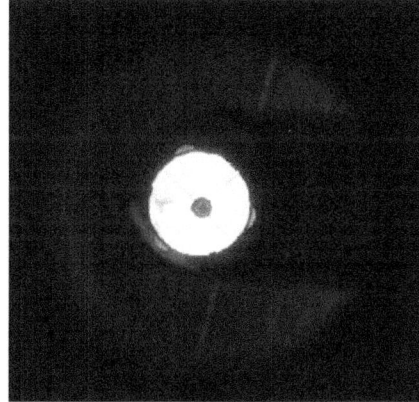

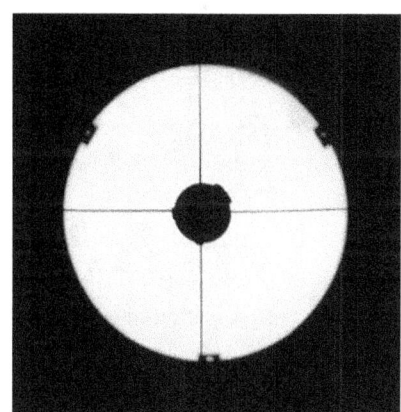

Figure 164   Figure 165

If we look directly down the center front end of the tube, we should see a view similar to that in Figure 166 below.

Figure 166

Okay, the telescope is ready to be used. Put the eyepiece giving the lowest magnification and widest field of view into the focuser drawtube and tighten the screw to hold in place. My collection of eyepieces is pictured below in Figure 167. Since this telescope has a focuser that takes U.S. standard 1 ¼ inch diameter barrel size eyepieces, there are two I like to use along with a 2.5X Barlow lens. For low power, I use a Tele Vue® 26mm Plössl eyepiece giving me about 60X. For higher magnification, I use a Brandon 16mm orthoscopic eyepiece from VERNONscope & Company, which gives about 100X. If I use the Tele Vue® 2.5X Barlow and insert the 26mm eyepiece, it now gives me 150X. With the 16mm eyepiece and Barlow, I now get 250X. This is an ideal range of magnifications to observe most all celestial objects well. Note Figure 168 below to see these two oculars and Barlow.

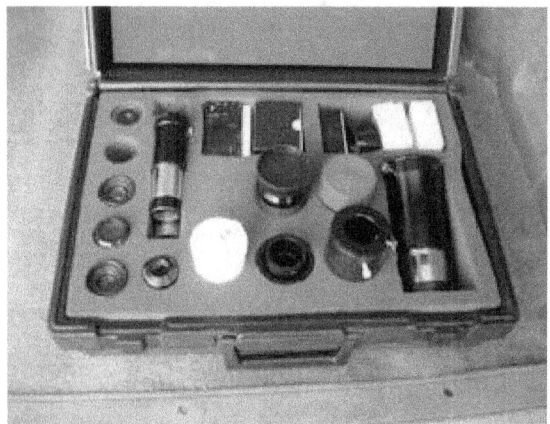

Figure 167

Figure 168

Before we aim our telescope to observe a particular object, let's test the range of motion for where it is capable of pointing in the sky. If we constructed all according to plan, then we should be able to point the telescope slightly beyond straight overhead (the point called the zenith), so that the bottom end ring and mirror cell hits against the front board of the rocker box after passing this high point, as seen in Figures 169 and 170 below.

Figure 169

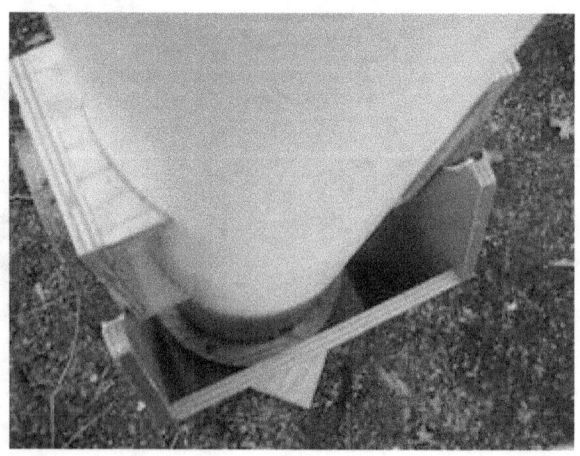

Figure 170

Another extreme measure is to be able to point the telescope down a bit below the horizon, so that the bottom of the tube does not hit against the top of the front board of the rocker box until after passing this low point, as seen in Figures 171 and 172 below.

Figure 171

Figure 172

Now aim the telescope to point at a terrestrial object such as a stop sign or light fixture. Get the object centered in the eyepiece's field of view. Then, look into the finder scope and tighten or loosen each of the 3 screws to collimate it so the same object is centered in the crosshairs.

One final check is that the optical tube assembly balances well and stays in position wherever we point it in the sky. A small sand bag can be used to help especially if the eyepiece is a bit heavier and needing to be counterbalanced. Friction clutches can also be installed to help with this. If the ground or pavement below is relatively level, we should have no problem with left to right motion in azimuth.

Well, now let's choose a celestial object to observe in the night sky, such as the moon or a planet like Mars, Jupiter or Saturn. First, aim the telescope tube like you were pointing a cannon, sighting along its length. Second, look into the finder scope and align the object so it appears centered in its crosshairs. Finally, using the lowest magnification, look into the eyepiece of the telescope and make small adjustments to center the image after focusing so the stars are sharp points of light or the object is sharply defined. Use this same exact procedure every time you want to look at another object.

The sheer pleasure of observing celestial objects may be something you want to share with everyone. For young children or shorter people, a folding step stool helps them reach the eyepiece, as seen in Figure 173.

Figure 173

We can bring our telescope to a daytime event and share the joy of telescope making, as I did at an annual Sea Fair by the water in Figure 174 below.

Figure 174

Or we can share a view of sunspots on the Sun by using a solar filter, as seen in Figure 175. The reason the filter looks so much larger than it needs to be is because I had this one already fitted for a 12-inch telescope I built. So instead of getting another one, I decided to retrofit it to work with the 8-inch telescope as well. You can imagine the challenge I had to counterbalance this assembly! ☺

Figure 175

You may wish to find out if there is an amateur astronomy organization in your area and look them up. There is a list of 888 astronomy groups (as of this writing) available online at http://astronomy.com. Both Astronomy and Sky & Telescope magazines are great resources to learn more about astronomy and telescope observing.

Up, Up & Away

So what can we see with our new and powerful telescope? Well, the sky's the limit is a cliché that does apply here, literally. Recently, we had what is called a Super Full Moon, which means that the closest approach the moon reaches the earth in its orbit (called perigee) coincides with the full moon phase. That's somewhat of a rare event in that it only happens once every 18 years or so on average. Color filters are available that screw on most eyepieces to reduce glare on the moon and bring out planetary features and details as well. Full moon, as seen in Figure 176 below, is not the best time to view the lunar landscape telescopically because many of the features are washed out by all the bright light from the sun. The best time to view the moon is when there is a distinct line, known as the terminator, which separates sunlit portions from shadow, presenting a large amount of contrast. One common time to do this is during or near the first quarter phase. Half the moon facing us is reflecting sunlight while the other half is dark facing away from the sun. The terminator would henceforth be right along the middle of the moon, showing some beautiful views of craters in the highlands area. See Figures 177-179 for such striking close-up views.

Figure 176

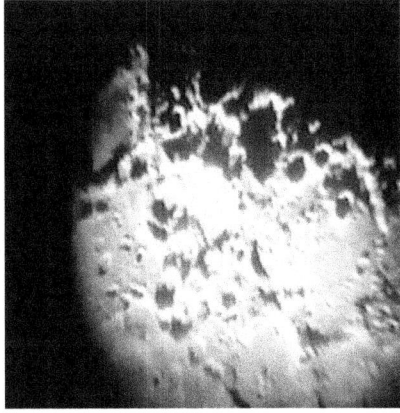

Figure 177

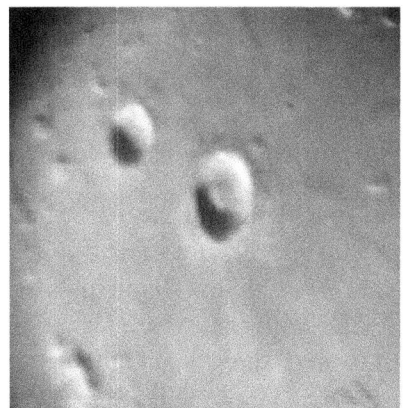

Figure 178

Figure 179

If you have an optical solar filter you can safely view sunspots on the sun, as viewed below in Figures 180 and 181 below.

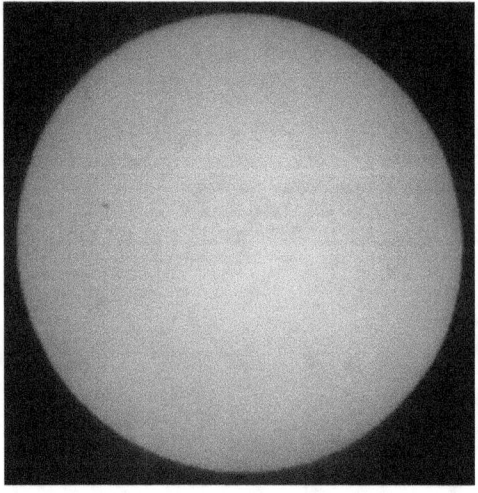

Figure 180

Figure 181

You can also witness an event where the planets Mercury or Venus pass in front of the sun from our line of site, called a transit. The rarer of the two is pictured below in Figures 182 and 183 when a transit of Venus was nearly over.

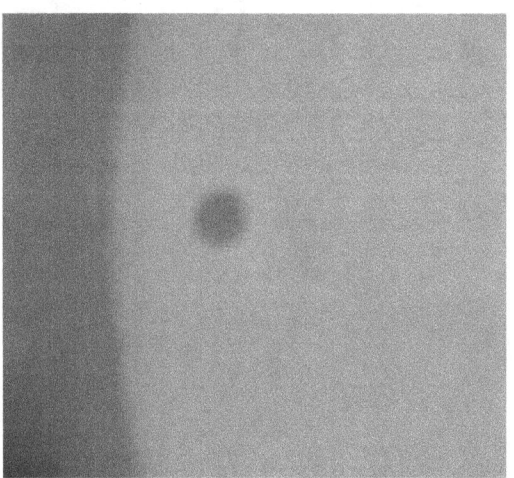

Figure 182

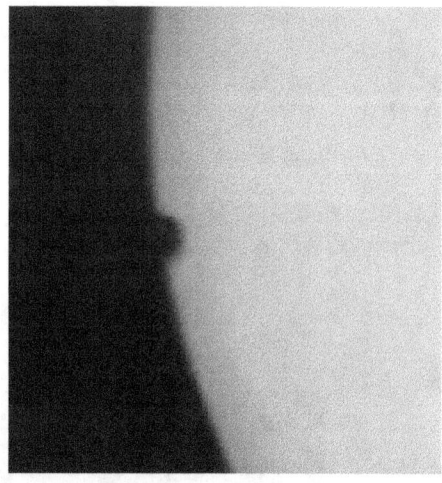

Figure 183

Other objects not requiring a solar filter that we can view in our telescope include the largest planet in the solar system, Jupiter. Jupiter is 11 times larger in diameter than the earth, and yet its day is less than half of ours, spinning in rotation in a mere 10 hours and 40 minutes. We can easily see the cloud belts produced from differential rotation of what's below, and the four Galilean satellites discovered 400 years ago named Io, Calisto, Europa and Ganymede. See some views of Jupiter in Figures 184 and 185.

 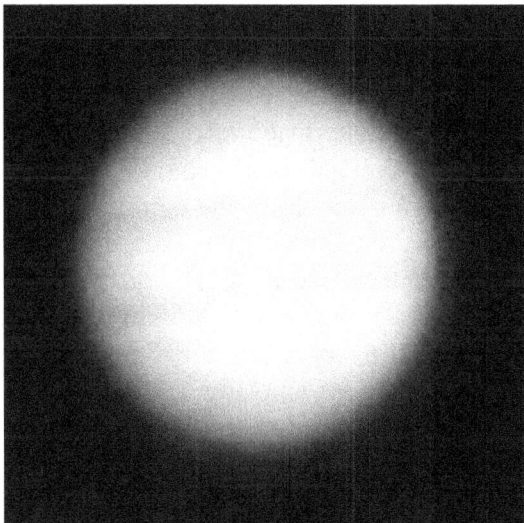

Figure 184   Figure 185

Saturn, with its prominent rings, is nearly twice as far away as Jupiter and the second largest planet in the solar system. Of Saturn's 63 moons, one of them named Titan is the largest moon in the entire solar system being nearly earth-sized and having an atmosphere. Its density is so light that Saturn would literally float if we could place it in the ocean. The rings, made of billions of icy rocks orbiting the planet, reflect sunlight as well and show up amazingly well in our telescope's view. As with Jupiter, 100X or higher will yield a good sized image with lots of detail visible, and does appear much sharper than the photo I took below in Figure 186.

Figure 186

Charles Messier was a French astronomer a few hundred years ago and like other scientists wanted to receive recognition for his work through making new discoveries. Messier was no exception to that rule being a voracious comet hunter, so he would spend many clear nights scanning the sky for these celestial interlopers. He found some, but he also bumped into many objects that resembled a cometary nucleus out in deep space, a fuzzy patch of light if you will, and found upon subsequent evenings of observation that these fuzzy patches did not move relative to the stars. That's a problem since comets do move, so Messier made a list of 110 of these "nuisances." What were these objects to us? Most all were galaxies, nebulae and star clusters. Messier #1, or M1 for short, is the famous Crab Nebula in the constellation Taurus the Bull. It is the remnant of a celestial explosion the Chinese first witnessed in 1054 AD. It was so bright that it could be seen during the day, and at night outshined the full moon. The remnant cloud seen today is what we see pictured below in Figure 187. Using the Chandra X-Ray Observatory launched into space on the Space Shuttle with the Harvard-Smithsonian Center for Astrophysics in Cambridge, MA around the turn of the millennium, an image of the stellar corpse remnant of this supernova was produced. Called a pulsar due to the "lighthouse effect" produced by the rapidly rotating neutron star of X-Ray emission, a teaspoon of this mountain-sized object would outweigh the Empire State Building in New York City.

Figure 187

The most distant naked-eye object in the night sky is another island universe containing some 100 billion stars, gas and dust about 2 million light years away. The Andromeda Galaxy, or M31, is quite similar in size and composition as our own Milky Way, and we are on a collision course with this behemoth set to take place billions of years henceforth. Some of the beautiful dark dust lanes within the spiral arms are revealed in the photo in Figure 188. Another galaxy known as the Whirlpool, or M51, not far from the handle of the Big Dipper in Canes Venatici, is shown in Figure 189.

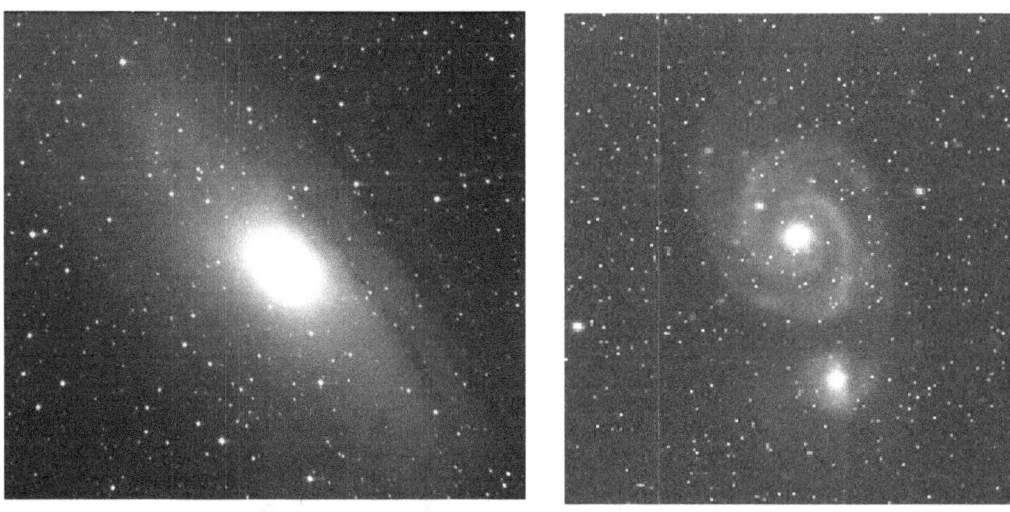

Figure 188   Figure 189

The famous Ring Nebula, M57 in Lyra, is another celestial showpiece in the heavens waiting to be rediscovered with our new telescope. Since we don't have setting circles on an equatorial mount, we can "star hop" our way easily to this object by looking between the two stars at the end of a slanted parallelogram figure of the Lyre, or harp. All celestial objects in the night sky can be found using this technique with a good set of star charts, a red flashlight to retain dark-adapted vision, and some patience. In the Reference section at the back of this book you will find a daily observation log along with instructions on filling out the data requested. This tool used in conjunction with a planisphere, or star finder, is a winning combination to becoming familiar with the night sky. There are other catalog listings of celestial objects besides those of Messier, such as an Index Catalogue (IC) and a New General Catalog (NGC) amongst others. NGC 253 is a galaxy in Sculptor that is relatively bright and certainly worth our time observing through our new telescope. Both of these objects are pictured below in Figures 190 and 191.

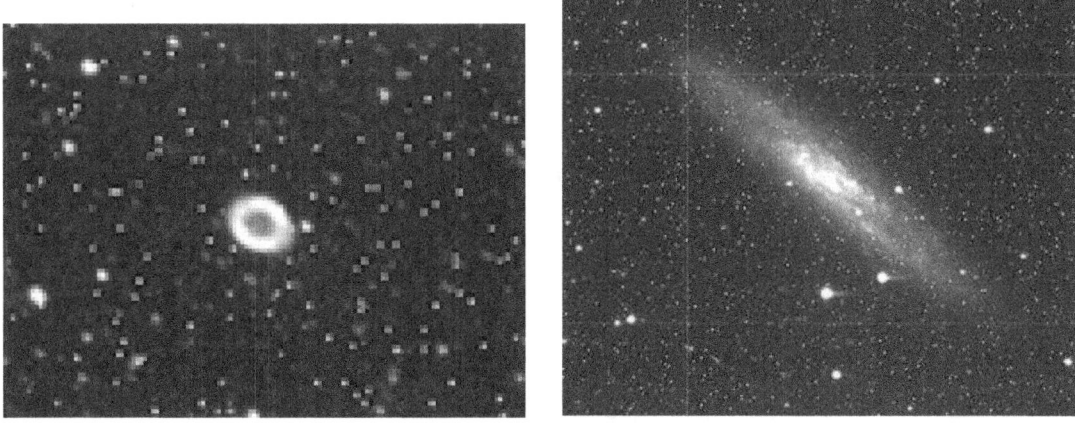

Figure 190   Figure 191

Two more objects of great interest in the winter sky include the Great Orion Nebula (M42) and the Pleiades Star Cluster (M45) in the shoulder of Taurus the Bull, otherwise commonly referred to as the Seven Sisters. The Orion Nebula is a diffuse cloud of gas and dust where new stars and planets are being born nearly 1,500 light years away. It is visible with the unaided eye in the middle of the sword, or dagger below the hunter's belt. The Pleiades are bright and easily visible as an open star cluster containing around 250 stars altogether. See their images below in Figures 192 and 193.

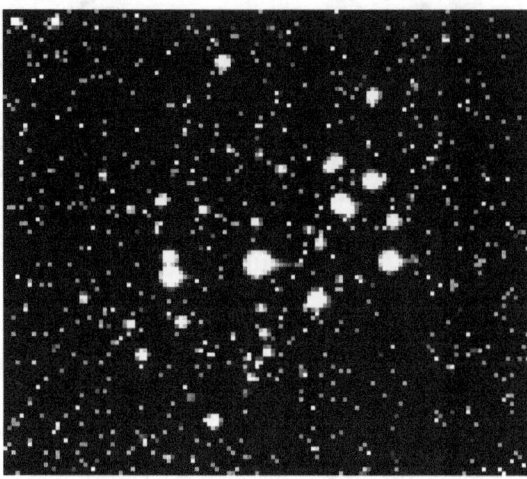

Figure 192   Figure 193

When not in use, always keep dust out of the telescope by covering the top front end with a clean plastic garbage bag. The common 13 gallon variety works perfectly well with our 8-inch telescope. Eventually, dust does accumulate upon the surface of our primary mirror and there is a method to clean it without scratching or harming its delicate optical surface. Materials needed include a basin large enough to easily fit the mirror into, surgical grade cotton, dishwashing liquid, and distilled water. To loosen the accumulated dust, we begin by immersing the mirror slowly into a bath of warm water in a basin. A 10-quart plastic dishpan can be purchased for around $2.00 and works well for this purpose, as seen below in Figures 194 and 195.

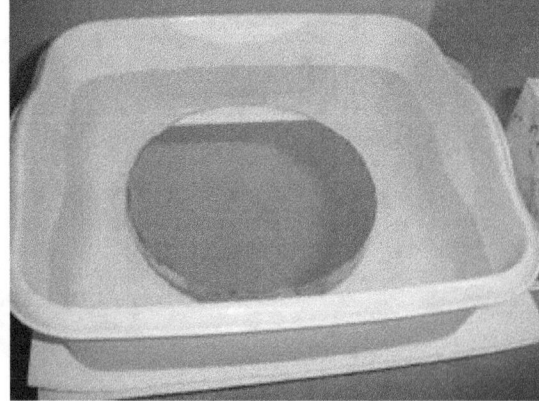

Figure 194   Figure 195

After allowing the mirror to soak a little while, remove the mirror and take a small handful of surgical grade cotton (regular cotton contains fibers that will scratch the mirror's surface) and soak under running warm water. Then put a few drops of mild dishwashing liquid on the cotton wad. See Figures 196 and 197 below.

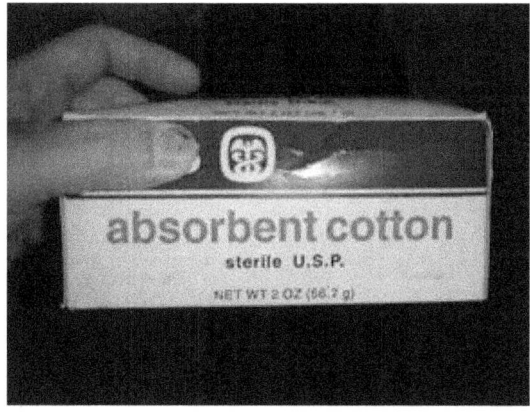

Figure 196

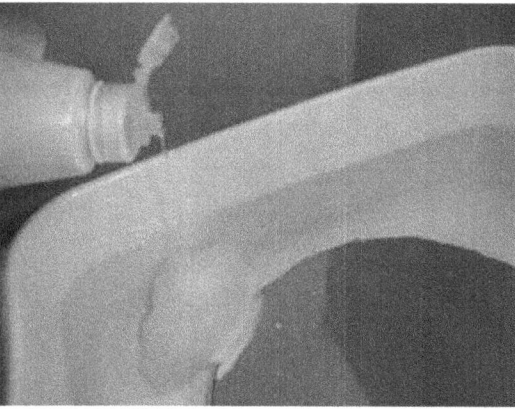

Figure 197

Now take the cotton and very gently rub across the face of the mirror back and forth, allowing only the weight of the wet cotton itself to swipe the mirror, as seen in Figures 198 and 199 below. A small amount of suds will form.

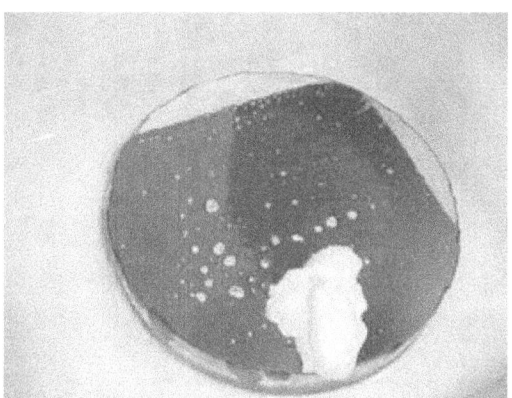

Figure 198

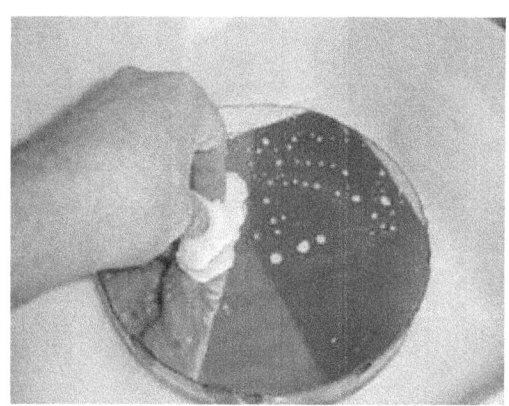

Figure 199

Now we will rinse the suds off the mirror under running warm water, as shown in Figure 200.

Figure 200

Do a final rinse using distilled water so it will air dry without leaving water spots behind as tap water would. See Figures 201 and 202 below.

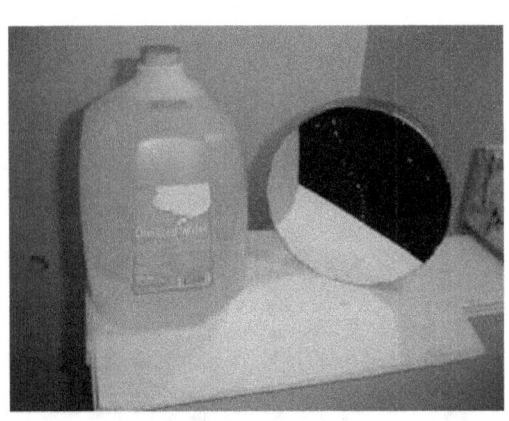

Figure 201

Figure 202

Set the mirror upright on its edge in a safe location where it won't fall and allow to air dry, as seen in Figures 203 and 204 below (which shows where NOT to place it).

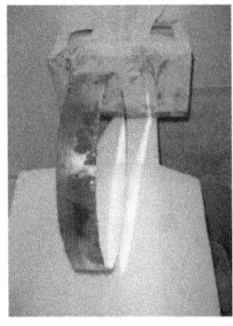

Figure 203

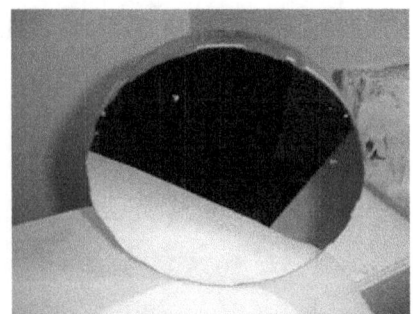

Figure 204

To transport the telescope, it fits easily into an SUV or van. For my Chevy Astrovan, I simply slide the tube from the back over the top of the seat backrests. It's important to secure the tube assembly using a bungee cord or rope to keep it from sliding too far forward. The rocker box and ground board easily fit on the back floor space long with an eyepiece case and other items to bring for an observing session, as seen in Figures 205 and 206 below.

Figure 205

Figure 206

For longer travels, I like to take the primary mirror out and put it on the floor under a seat in some container. If you don't have a box or something else, standard 10 ¼ inch wide microwavable paper plates placed above and below the mirror gives some form of protection from dirt or debris, as seen in Figures 207 and 208 below.

Figure 207

Figure 208

Through a little care and maintenance, our new telescope should serve us well for many years to come. Indeed, other than the need to recoat the surface of the mirror after several years of use, and perhaps a little touch up paint now and then, our telescope can easily last us a lifetime and beyond.

References and Helpful Information

Telescope Primary Mirrors

Discovery Telescopes, Inc.
P.O. Box 1278
Stanton, CA 90680
http://www.discoverytelescope.com/optics.html
Order By Phone 1-877-523-4400 or 949-347-0142
Bill Larsen

Mirror Kits

Firsthand Discovery, LLC
P.O. Box 1724
Versailles, KY 40383
http://firsthanddiscovery.com/8-inch-telescope-mirror-kit.html

Mirror Starter Kits

Newport Glass Works, LTD.
P.O. Box 127
Stanton, CA 90680
Tel: (714)484-7500, (714)484-8100
Fax: (714)484-7600 or (714)484-8181
E-mail: newportglass.sales@gmail.com
http://www.newportglass.com/amrrkit.htm

Telescope Making Supplies

Willmann-Bell, Inc.
P.O. Box 35025
Richmond, Virginia 23235 USA
http://www.willbell.com/ATMSupplies/ATM_Supplies.htm
Phone: (804) 320-7016
Toll-Free: 1-800-825-STAR (7827)
Fax: (804) 272-5920

Teflon Bearing Kits

AstroSystems Pivot Kit 3/8" Bolt $27.00 Improves Telescope Motion Each kit contains a 3/8" or 1/2" x 2.5" brass bolt, bronze thrust bushing, stainless steel locknut assembly and full instructions. Teflon Bearing Kits - Small kit $22.00. All kits contain 3 azimuth and 4 altitude bearing pads of 3/16" minimum thickness virgin Teflon, pre-drilled and countersunk, with stainless steel mounting screws. See http://www.astrosystems.biz/pivot.htm

ScopeStuff
PO Box 3754
Cedar Park, TX 78630-3754
Email: support@scopestuff.com
512-259-9778
http://www.scopestuff.com/ss_lexx.htm
$17.00 Four Virgin White PTFE Pads, 1/8" Thick, 3/4" x 1", Drilled and Countersunk, With Screws
$12.00 Ebony Star Laminate Strip, 3/4" x 48"
$13.00 Ebony Star Laminate Strip, 1" x 48"
$32.00 Ebony Star Laminate Disk, 16" OD, No Center Hole, No Pads (ask if will sell 13.5" OD disk)
$42.00 Ebony Star Laminate Round Bearing Kit, 13.5" OD, 1.25" Width, With PTFE Pads
Each of the round bearings comes with three PTFE pads, 1 inch square and 1/8" thick, which attach with contact cement and supplied stainless steel wood screws.

Parks Optical also has some of the best telescope mirrors available.
http://www.parksoptical.com/

Stellafane Links Page has a large list of links to almost everything required to build a telescope.
http://stellafane.org/misc/links.html

Daily Observation Log

Observer: _____   **Date:** _____

Time: _____ am / pm   **Duration:** _____min

Sky: 0 1 2 3 4 5 (circle one)   **Seeing:**   0 1 2 3 4 5 (circle one)

Constellation(s): _____

_____

_____

_____

Star(s): _____

_____

_____

Planet(s): _____

Object(s): _____

Phenomena: _____

Observational Method: unaided eye   paper tube   binoculars   telescope   (circle one)

Drawing:

Daily Observation Log

Observer: _____ **Date:** _____

Time: _____ am/pm **Duration:** _____min

Sky: 0 1 2 3 4 5 (circle one) **Seeing:** 0 1 2 3 4 5 (circle one)

Constellation(s): _____

_____

_____

_____

Star(s): _____

_____

_____

Planet(s): _____

Object(s): _____

Phenomena: _____

Observational Method: unaided eye paper tube binoculars telescope (circle one)

Drawing:

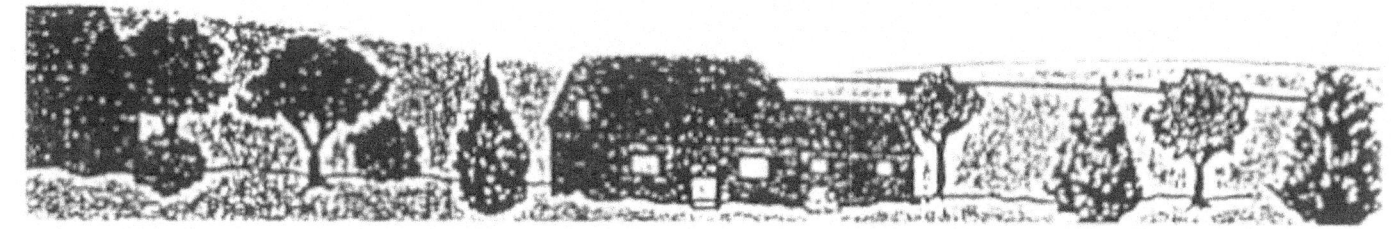

Instructions for Completing Daily Observation Log

Observer: Please print your full name

Date: Record current month/day/year (i.e. 01/08/2009)

Time: Record the time you began the observation and circle AM or PM

Duration: Record the total number of minutes you actually made your observation

Sky: Circle one number that best represents the sky from clear to completely overcast. 0 = clear; 1 = a few small clouds; 2 = partly cloudy; 3 = sky 50% cloud-covered; 4 = few breaks in clouds; 5 = completely overcast

Seeing: Circle one number that best represents the seeing conditions from excellent to poor. "Seeing" is a term used by astronomers to describe the steadiness of the atmosphere. One method of determining how steady or unsteady the atmosphere is, due to air currents and temperature changes, is by studying the brighter stars.
Bright stars that appear to "twinkle" indicate turbulence in the layers of air in the atmosphere. Rate the seeing conditions on a scale of 0 for perfectly steady to 5 for stars that appear to "dance" in the sky.

Constellation(s): List any constellation you are able to identify in the night sky.

Star(s): Write the name of each brightest star you are able to identify by consulting a star chart or atlas.

Planet(s): Write the name of any planet you identify by referring to current data available giving its location.

Object(s): Record the number and types of objects seen in the sky. Examples include meteors ("falling or shooting stars"), satellites, comets, asteroids, etc.

Phenomena: Any form of sky glow, such as aurora or the Milky Way, may be recorded

Observational Method: Circle the method of observation used. More than one per observation period can be utilized.

Drawing: Draw the moon phase (amount of sunlit portion) if visible. Also draw in anything recorded for that day's observation. You should draw in boundary lines separating different parts of the sky and include the direction abbreviated (i.e. SW) for each segment.

Kevin Manning is an international award winning astronomer, and while he is supposed to be retired, he's as busy as ever doing a nationwide Star Tour! Kevin won national and international awards in his field, and did some work with Brookhaven National Laboratory. Kevin was both a Wright Fellow at Tufts University and an Einstein Fellow working on Capitol Hill in Washington, DC. He served as an Editor for the U.S. Department of Energy's Office of Science Journal of Undergraduate Research, 2001 (vol. 2), and worked with the International Atomic Energy Agency (IAEA) with the U.S. Support Program (USSP). Besides the numerous workshops he's presented over the years at libraries, observatories, and science centers, some noteworthy ones include those made at Tufts University, State University of New York at Stony Brook, the NSTA National Convention, AAAS Breakfast with Scientists, and the National Parks Service. While Kevin's scientific background is stellar and his mind is definitely in the stars, he's also one of the most down to earth and accessible guys you'll ever have the chance to meet. Kevin goes to great lengths to communicate his passion for the amazing world of outer space in terms that all ages can understand. He's lectured widely, built telescopes and observatories, and consulted with NASA, but his favorite thing to do is to share his contagious enthusiasm and knowledge with the public. Astronomy is more than a study for Kevin, it's his love and it shows in all he does.